KB068496

THE
SCIENCE
OF
OLYMPUS

This book is dedicated to my family

And my teacher Peter,

Who started me on the path to Enlightenment

"If something imaginary is widely believed to be real, is it still imaginary?"

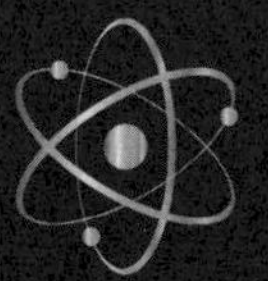

THE SCIENCE OF OLYMPUS

• written by •

Jay Joonyoung Chang

바른북스

/ Introduction /

I first touched Greek Mythology at the age of seven and a half. Back then, it was in the form of comic books featuring the Greek myths bought for me by my mother. I spared them no interest at first. They were just a book series amongst many, collecting dust on my bookshelf.

Then, when the rainy season came, I found myself having nothing to do after school. I couldn't go play outside with my friends, my parents were out at work, and I didn't want to read my favorite books that I read over what felt like a hundred times.

Overcome with boredom and lethargy, I flung myself down on my bed and stared up at my ceiling and my bookcases. Then, I found my attention being drawn to a comic book series so long that it was taking up an entire shelf.

They were the Greek myth comic books. And because they looked vaguely interesting, I remember thinking to myself 'Meh, maybe I'll give them a read'. It was one of the best decisions of my life.

As soon as I read the first few pages, I was immersed in a world of gods, monsters, and heroes. After finishing the first few books ravenously, I thought 'Why didn't I read these books before? All this time, these amazing books were just sitting on my bookshelf! How come I didn't notice them before?'.

When my parents came home, my mother was pleased to see me reading the myths she had wanted me to read long before. I regret not heeding her words. When she told me that I would like the books, I had just replied that they were long and looked boring. What I experienced that day cannot be described. I had entered a world where everything was possible.

When I was eight, my father got a chance at his job

to go to America. My family unanimously agreed that we would go.

It turned out that my family was going to live in Virginia, Fairfax County. I was terribly excited to finally go to America after dreaming about it for the past few months and when our plane touched down at the airport, I couldn't stop smiling.

My twin sister and I were enrolled in Westbriar Elementary School. It wasn't until about a year after we came to America in the third grade when I finally met my best friend, William. Our connection? Greek mythology.

Like me, William loved books with a passion equal to or perhaps more than my passion. He introduced me to a Greek mythology book series and I was hooked. The book series was mind-blowing. It was even more enjoyable than the comic books from which I found the special world of Greek myths.

Also, after I finally took the end term exams in the last semester of first grade and the first semester of second grade, I took solace in the books to recover and prepare for the next exams.

It was in these times that I was able to move onwards to more and more advanced books of Greek mythology and read the Iliad and the Odyssey by Homer along with Mythology by Edith Hamilton.

In the summer of 7th grade, I traveled to Greece and the ruins I had longed to visit when I was younger.

Greek mythology was the ancient Greeks' science. Science is a collection of ideas and facts which we believe is true. In the ancient world, to these people, the gods were an integral part of their life. They were real.

I wrote this book with the intention of telling the science of a world long past, lost to the sands of time, along with the science of today.

By beginning each chapter with myths about a particular god, and ending with an interpretation of each myth with the science of today, this book is designed to kindle an interest in science.

The science concerned with the myths begin in Part 2: Reign of Olympus, while Part One: Before the Gods lays out the background for the myths to take place in. Part One will also include my own opinion on the myths.

/ contents /

Part One : Before the Gods

Part Two : Reign of Olympus

01

Part One :

Before the Gods

First there was Chaos, the vast immeasurable abyss,
Outrageous as a sea, dark, wasteful, wild.

Chapter One :

The Beginning

The Beginning

Earth, the beautiful, rose up,

Broad-bosomed, she that is the steadfast base

Of all things. And fair Earth first bore

The starry Heaven, equal to herself,

To cover her on all sides and to be

A home forever for the blessed gods.

01

The Gods were not the first all-powerful beings to rule over the earth. Before them, there was their birth race, the Titans, and before even them, the Primordials.

In the beginning, there was Chaos, the first. Chaos gave birth to Gaea, goddess of the Earth, and she, in turn, gave birth to the sky and ocean, Ouranos and Pontus respectively. Gaea and Ouranos then married and gave birth to three races: the Titans, the Cyclopes, and the Hecatoncheires.

The Titans were humanoid in shape but like the gods, were powerful deities that could change the world and nature to their whims. There were twelve of them in total. Six brothers and six sisters. The brothers included

Oceanus, Koios, Krios, Hyperion, Iapetus, and Kronos. The sisters were Theia, Rhea, Themis, Mnemosyne, Phoebe, and Tethys.

The Cyclopes were a race of single-eyed monsters who enjoyed metalworking and the Hecatoncheires were the Hundred Handed Ones, literally meaning that they had a hundred hands. There were three of them and their names were Cottus, Briareus, and Gyges.

Ouranos was pleased with his firstborn race, the Titans, but was angry with the Cyclopes and the Hecatoncheires because they looked like monsters, and could not believe such creatures had come of his body.

Ashamed of how his second and third children looked, Ouranos thrust them down into Tartarus, the abyss below the earth which held all evil things and was made by Chaos as a reverse of the sky.

Gaea was unhappy with her husband throwing her children away into the Earth. Although they looked much like monsters, they were still her children and a mother's love for her children holds all-powerful against all other odds.

She then reached deep inside the Earth and created the first weapon: the scythe. Power-hungry, the last born of the Titans, Kronos seized the weapon and volunteered to help his mother obliterate his father.

Kronos enlisted the help of four of his brothers, Krios, Koios, Iapetus, and Hyperion, and promised them the four corners of the earth if they succeeded in overthrowing their father and seizing his throne.

Gaea tricked Ouranos, and Kronos cut his father up with the scythe and tossed his body pieces into the air while his four brothers held Ouranos down. The blood of Ouranos got everywhere and it held life.

When it met harsher forms of nature such as rocks and boulders, the blood transformed into sturdy lifeforms or monsters. In fact, the three Furies, the helpers of Hades, god of the Underworld, were born this way.

If the blood touched softer lifeforms, like trees or rivers, they each sprouted a nature spirit. Tree spirits were dryads and river spirits were naiads and so on. They were minor nature gods.

In Greek mythology, the beginning of the universe is

strange and cruel. Ouranos throwing his children back into the earth just because of what they looked like is a cruel thing for anyone to experience from another person, let alone his father.

This scene of throwing them into the earth and nobody doing anything about it looks like our society today treating the social minority. People know that they are there but are ashamed of them and pretend like they don't exist or make fun of them.

I also want to point out the actions of Gaea. I understand that she was angry and shocked at her husband Ouranos for throwing their children into Tartarus because although they looked like monsters, they were still her children and her love for them crossed all boundaries.

In the news, you sometimes see freak accidents about mothers gaining the super human strength to save her children.

This is the power of love a parent has for their child. Gaea was the same. Her love for her children enabled her to sacrifice her husband for her children's wellbeing. As you will see in the later chapters, Gaea's primary aim was to love her children and have the best for them.

I believe that this is true in every mother. They would all work for and even sacrifice themselves for their children. In this way, all mothers in the world are true heroes who made us who we are now.

If circumstances were different, Ouranos could have been a good father for his children too. No one, not even the gods can have such hatred for their children. In my perspective, being the master of the universe was a bit too much for him and he was overcome with the stress and overwork which caused him to snap.

In Korean society today, many parents are overworked like Ouranos. They all work late into the night to give their children the best education and to enable them with skills to succeed in the world. So, I look at Ouranos not as a figure of hatred and cruelty but a figure of sympathy and a person that can develop.

Chapter Two :

The Golden Age

"No man or woman born, coward or brave, can shun his destiny."

When Ouranos was defeated, it was the beginning of a new age, the age of Titans, or its official name which is the Golden Age. The six brothers, Oceanus, Koios, Krios, Hyperion, Iapetus, and Kronos along with the six sisters Theia, Rhea, Themis, Mnemosyne, Phoebe, and Tethys all chose their respective domains.

Oceanus was the Titan of the sea and Pontus willingly gave his throne to him. Because Koios, Krios, Hyperion, and Iapetus helped in Kronos's murder of Ouranos, they each got the four corners of the earth which were North, South, East, and West respectively. Koios was the Titan of the North and had an insight into the future. Krios was the Titan of the South, Hyperion, the Titan of the East, Iapetus the Titan of the West, and lastly, Kronos was the ruler of the universe along with being the Titan of Time.

The sisters were Theia, the shining one, Rhea the Great Mother, Themis, Titaness of justice, Mnemosyne Titaness of memory, Phoebe, Titaness of foresight, and Tethys, Titaness of fresh water. The Cyclopes and the Hecatoncheires built them a palace on the highest mountain in Greece: Mount Othrys.

To keep the power only within the Titans, Kronos cast his siblings, the Cyclopes and the Hecatoncheires back down into Tartarus. In order to further strengthen their power, the Titans all intermarried each other to keep the power in their family except for Kronos.

The reason for that was because when Ouranos died, he had issued a prophecy about Kronos that he was destined to be overcome by his own children, just as he had overthrown his father.

However, Kronos had children with Rhea and sired six of the twelve Olympians. However, when his first child, Hestia was born, he turned out to be afraid of his father's prophecy and decided to prevent it by eating his children. Five of his children, Hestia, Demeter, Hera, Hades, and Poseidon in birth order, were swallowed in this way.

However, Rhea, sick of her husband's actions, called out to her mother Gaea for help and advice. Gaea advised her to go to Crete and give birth to Zeus, her last child, there. A goat named Amalthea along with some nymphs agreed to raise Zeus, and Gaea sent some soldiers named the Kouretes to clash their shields and swords together so that Kronos would not hear Zeus crying from his palace of Mount Othrys. As for Rhea, she took a small rock, swaddled it in baby blankets, and presented it to Kronos. Kronos swallowed the disguised rock without hesitation.

When Zeus came of age, he disguised himself as one of the Titans and became Kronos's cupbearer. One night, he mixed Kronos's wine with mustard seeds, which made him puke all the things out of his stomach. Zeus's siblings came out of Kronos's gullet in the reverse order that they were born (Poseidon, Hades, Hera, Demeter, and Hestia, the order from which they were disgorged from Kronos's stomach and now fully grown).

Ouranos's prophecy was a self-fulfilling one. The

prophecy is ironic because Kronos was so afraid of the prophecy his father had issued him that he went to great lengths to prevent it, which in turn, made it come true.

So, in the Greek mythology world, I think that one's fate was unavoidable, only delayed. This kind of fight in the storyline is the kind of fight where the main character defies his fate.

In the story of the defeat of Kronos, Kronos tries not to give into his fate and acts to prevent it. In Korea, we have a similar type of story called <The Post Horse> or in Korean, 역마(Yong-ma).

In the novel, the main character has itchy feet and is destined to spend his life wandering around. To prevent her son's fate, his mother tries all sorts of things and she thought she had succeeded when he fell in love, but it turned out that the woman was his aunt. So, the main character accepts his fate.

The similarity between these two stories is that the character's fate comes true no matter what. However, the difference is that in the

case of Kronos, he made the prophecy come true with the act of trying to prevent it, not just passively accepting his fate, while in 역마 (Yong-ma), the character accepts his fate.

When I read <The Post Horse> and the myth of Kronos, I thought that knowing the future is both a beautiful and terrible thing and that we humans should not attempt or presume to know the future or try to prevent it, even if there are truly three Fates weaving their loom on Olympus. The best that we can do is to live in the present and try our best to live life to the fullest.

Chapter Three :

The Titanomachy

A dreadful sound troubled the boundless sea.
The whole earth uttered a great cry.
Wide heaven, shaken, groaned.
From its foundation far Olympus reeled
Beneath the onrush of the deathless gods,
And trembling seized upon black Tartarus.

01

After Zeus freed his siblings, they all agreed to wage a war against their father. However, before they could do that, they needed weapons. Since he had thrown her children, the Cyclopes, and the Hecatoncheires back into Tartarus, Gaea now sided with Zeus and the other gods against her son Kronos.

Gaea advised them to go to her children in the underworld and to acquire weapons from them, as they had spent all this time honing their metal skills while being watched by a jailor by the name of Kampe. Zeus and his sibling allies freed the Cyclopes and the Hecatoncheires and in gratitude, they decided to side with Zeus and to build powerful weapons for their side such as Zeus's iconic thunderbolt, Poseidon's trident, and Hades's helm.

The Gods chose their base as Mount Olympus, the second-highest mountain after Mount Othrys. The Gods and the Titans then started a war that would last ten very long years called the Titanomachy.

The six siblings born of Kronos and Rhea were on one side along with the Cyclopes, the Hecatoncheires, and some 'good' Titans such as Prometheus and Themis. On the other side were Kronos, Hyperion, Atlas, and the rest of the male Titans. The female Titans remained neutral.

During the war, the Cyclopes made weapons for the Gods and their allies. The Hecatoncheires threw boulders with their one hundred hands. The end of the long war was when the gods went all out in a massive attack and destroyed the Titan base, therefore, incapacitating the Titans. After the dust had settled, it turned out that they had sheared off nearly half of Mount Othrys, making Olympus the highest mountain in Greece. The Gods had won.

They put the defeated Titans in Tartarus and made Atlas, the son of Iapetus, who used to boast about his strength to hold the sky away from the earth. The four Titans of the four corners of the Earth had to be there in order to keep the balance between the Heavens which was constantly straining to embrace the Earth. With them thrown into Tartarus, someone had to hold the Sky and keep it from resuming its primordial embrace with the Earth. The female Titans who were neutral and the Titans who had helped the Gods (Prometheus and Themis) in the war were spared from the punishment.

As they say, history repeats itself. The ruler makes a mistake with a mother and her children and the mother helps the opponent of the ruler rise to power and the ruler is punished. I think what the world was looking for back then was a ruler who wasn't greedy or power-hungry and ruled with justice.

The moral in this story was that everyone has a special part to play and each person has their own unique talents. So, never underestimate someone else.

An example of this in the Titanomachy was that Kronos didn't realize the worth and specialty the Cyclopes and the Hecatoncheires possessed and underestimated them, unlike Zeus who listened to Gaea's advice and brought them on his side.

The second moral of the story in my point of view was that it stated and reinforced the "Golden Rule"- to treat others like how you want to be treated, except in this case, it is true. Kronos and the other Titans tossed the Cyclopes and the Hecatoncheires into Tartarus but, later on, the scales were tipped and they threw Kronos and the other Titans into Tartarus instead.

Another lesson of the story was that in order to lead other people, you have to figure out what the people you want to lead want and try your best to give everyone what they want.

If Kronos had shown the female Titans a perspective of him that he was a fair and just leader, perhaps more of the female Titans would have joined him in war, therefore tipping the scale in his favor.

02

Part Two :

Reign of Olympus

Strange clouded fragments of an ancient glory,
Late lingerers of the company divine,
They breathe of that far world wherefrom they come,
Lost halls of heaven and Olympian air.

Chapter Four :

Zeus

I am mightiest of all.
Make trial that you may know.
Fasten a rope of gold to heaven and lay hold,
Every god and goddess.
You could not drag down Zeus.
But if I wished to drag you down, then I would.
The rope I would bind to a pinnacle of Olympus
And all would hang in air, yes, the very earth
And the sea too.

After Zeus and his six siblings won the war against the Titans, the Cyclopes and the Hecatoncheires worked together to build a magnificent palace on the gods base during the war: Mount Olympus.

Olympus, the palace of the gods, was said to be at the top of the mountain and was so high that it took nine days to fall from Olympus to the earth.

The gods lived in harmony on Mount Olympus, eating ambrosia and drinking nectar, which was the food and drink of the gods. The three brothers, Zeus, Poseidon, and Hades rolled dice to decide who got to choose first between the heavens, the sea, and the underworld. They agreed that the earth was neutral territory as it was between the three realms.

Zeus got the highest dice roll and he chose the heavens, Poseidon was the second-highest and chose the sea, and Hades was left with the underworld. Hera chose to be the goddess of marriage and family because of her marriage to Zeus, while Demeter decided to be the goddess of agriculture and Hestia elected to be the goddess of the family hearth.

However, Hestia wasn't one of the twelve Olympians because she would willingly give her throne on Olympus to Dionysus in order to avoid a fight later on in the future.

The one thing that I wanted to focus on in this chapter was that the female gods didn't join, or, weren't allowed to join the dice game, which was very unfair. This was because of the harsh gender discrimination that existed in the past.

In school, we learned that only male citizens of age were able to vote in Athens. Women couldn't vote. We know female discrimination existed, causing the Greeks to, even in their myths, assume that women have no power and want to take the backstage.

If one of Hera, Demeter, or Hestia took control over one of the three domains, there could have been much smoother proceedings. However, Zeus and Poseidon would have sired much fewer heroes, which would result in an ancient world having much more monsters than before. So, I guess that every cloud has a silver lining.

Zeus got the highest die roll when playing with his brothers and he chose the sky which was fitting because of the lightning bolt that the Cyclopes and the Hecatoncheires made for him during the Titanomachy. He was the god of the sky, justice, and order. His symbol was the lightning bolt (his main weapon) and his sacred animal was the eagle.

Zeus married Hera by turning himself into a small cuckoo bird and flying to her in the middle of the storm which ultimately convinced Hera to become his wife. Although Zeus was married to Hera, he sired a number of heroes with mortal women and even gave birth to new gods with other deities. So, much of the myths surrounding Zeus are not his heroic acts but of his love life.

For example, he took the form of a sweet, peaceful cow and got the princess Europa to sit on his back. He then crossed the ocean all the way to Crete where she was guarded by Talos made by Hephaestus from invaders and pirates.

Europa then gave birth to Minos who would later become the infamous king of Crete along with his two siblings, Rhadamanthys and Sarpedon. The ancient Greeks never knew where Europa was taken and so named the far-off continent Europe after Europa.

Another story was how Zeus sired the hero Perseus with Danae. Danae's parents, the king and queen of Argos, did not have a son after all their years of trying but were given a prophecy concerning their daughter, Danae.

The prophecy said that she would have a child and when the child grew up, it would kill its grandfather. The king of Argos was shocked and he immediately stopped the flow of suiters coming to court Danae and locked her up in a bronze tower.

However, Danae did not escape the notice of Zeus. He appeared in the form of a golden shower of rain and Danae became pregnant. She gave birth to the hero Perseus who would indeed, after a series of quests, kill the king of

Argos, his grandfather, by accident.

Although Zeus had many affairs behind his wife's back, he was also the god of law and order and punished the people who had committed crimes.

For example, the Titan Prometheus, one of the good Titans who supported Zeus in the Titanomachy made us humans and took pity on our existence and decided to sneak out the fire, which was only available to the gods back then and was one of the things that separated the gods from humans.

Prometheus stole the fire which was kept on Olympus in Hestia's hearth and snuck a few coals out in a licorice plant. (Some stories say that Hestia helped him steal the coals because she felt sorry for the humans too.) When Zeus found out about Prometheus's deception, he was not happy and chained him to a rock with an eagle eating his liver out every single day.

One time, Zeus decided that the whole human race was evil and he wanted to wipe the human race out. With the help of his brother Poseidon, he flooded the entire world.

However, Deucalion, who was the son of Prometheus and the Greek equivalent of Noah, received a message from his father and so constructed a boat. Deucalion, along with his wife Pyrrha, survived the flood.

Deucalion and Pyrrha prayed to Zeus and he allowed them to get a prophecy from the Oracle of Delphi in order to find a way to bring humans back. The Oracle of Delphi had incidentally survived the flood thanks to the temple of Apollo, the god of prophecy, which was located on the highest mountains.

To Deucalion and Pyrrha, the Oracle advised that to repopulate the earth, the two needed to throw their mother's bones behind their backs as they headed down from Delphi.

Deucalion and Pyrrha deduced that the Oracle must have meant not their literal mothers but the bones of the Mother of Earth, Gaea. Because the bones of the Earth were rocks, just like the Oracle advised, they throw rocks behind their backs as they descended from Delphi.

The rocks that were thrown from Deucalion's hand changed into men as they hit the ground and the rocks that left Pyrrha's changed into women. In this way, Deucalion and Pyrrha repopulated the Earth.

So, Zeus allowed the human race to repopulate the earth. This story goes to show that while Zeus is harsh, he can be merciful to the people who deserve his mercy and kindness.

Another example of this was when Zeus punished the king of Salmonea, Salmoneus. Salmoneus decided to imitate Zeus so that his citizens would respect him. He did not believe in the Greek Gods and thought that they were fake stories. If he had known they were real, he would not have done so. Salmoneus and his advisor pulled off the trick but when Zeus himself found out about Salmoneus's trickery; he destroyed the city into rubble.

It wasn't after I read these myths a couple of times that I actually understood the meaning behind the myths. This myth was made and used as a warning not to doubt the myths and to have trust and faith in the gods.

Also, it reinforces the fact that the gods are kind to good and pious people and rain punishment down on the people who deserve it. So, it seems that the ancient Greeks may have led a very conscious aware life.

Also, Zeus's cheating on his wife defines one of the most important concepts in biology today: Evolution. Or, to elaborate, natural selection and the survival of the fittest.

Evolution occurs when a change in the surrounding environment of an organism happens. To understand the concept of evolution, we first need to have a thorough understanding of genes.

Genes are how species that use sexual reproduction survive and continue into the next generation. They are what makes you look like your parents and the rest of your family.

Genes were discovered by the biologist, Gregor Mendel. Until Mendel, the scientific community had an idea of what genes were (something passed from the parents to the offspring which defines the offspring's traits) but what they did not know was what and how exactly that happened.

Then, Mendel conducted his famous experiment with peas. To do so, he decided that he would experiment on smooth peas versus wrinkled peas. Because there are only two types of green peas, smooth and wrinkled, this was an acceptable topic to conduct an experiment on.

The second stage was to grow and raise the peas so that no smooth pea would have a wrinkled pea gene and no wrinkled pea would have a smooth pea gene. To put it short, Mendel would raise two purebred lines, one of smooth peas and the other of wrinkled ones.

In order to achieve this goal, Mendel separated the peas into two groups: smooth and wrinkled. Then, he kept cross-pollinating between peas from the same group that they themselves came from. This cross-pollinating between the same group went on for many generations.

Finally, when Mendel was sure that he had a purebred line of smooth peas and a purebred line of wrinkled peas, he then cross-pollinated a pure smooth pea and a pure round pea. Mendel did this for many of the peas. And then, what he discovered from the results was shocking.

The offspring of a smooth and wrinkled pea had a four in one chance of being wrinkled while having a four in three chance of being smooth.

The next stage of his experiment was to self-pollinate all the offspring of the pure smooth peas and the pure wrinkled ones.

From his experiments, Mendel observed the following facts.

- The genes passed from the parents to the offspring exist in pairs, and the genes he named alleles.
- When the two genes that make up the organism are the same, he called them homozygous. When they were different, they were called heterozygous.
- There were dominant genes and there were recessive genes.
- If there was a dominant gene in any of the two genes deciding the trait of the organism, the pea would have the dominant trait. However, the two genes would need to be both recessive genes for the pea to have a recessive trait.
- The dominant gene is depicted with a capital letter, such as T, while the recessive gene is depicted with one of a lowercase letter, such as t.
- There were phenotypes, where one gene would be a dominant gene and the other would be a recessive

gene. In this case, because one of the genes making up the organism is a dominant one, the plant would therefore have the dominant trait. However, it would still have a recessive gene that it could pass on to its offspring.

From these observations, Mendel proposed three laws. The law of dominance, the law of segregation, and the law of independent assortment. The law of dominance states that the allele whose characters are expressed over the other allele is the dominant allele and the characters of this dominant allele are called dominant characters.

The second law, the law of segregation, says that when two traits come together in one hybrid pair, the two characters do not mix, but are independent of one another. The organism receives one of the two alleles from each parent.

To illustrate the law of segregation, let's say we have two peas, both a phenotype. Then they would both be expressed as Tt and Tt. The offspring of the two peas would have a 25% chance of being TT, a 50% chance of being Tt, and a 25% chance of being tt.

This is due to both of the parents passing on only one of their genes. The entire number of cases would be TT, Tt, tT, and tt.

The third and last law, the law of independent assortment, states that at the time of meiosis, there are separate genes for separate traits and characters and they influence and sort themselves independently of the other genes making up the organism.

To elaborate further, the genes that decide whether a pea is smooth or wrinkled, do not have any influence on whether the pea is green or yellow.

Let's get back to the original topic at hand, evolution. The concept of evolution was first defined by Charles Darwin. Darwin noticed the occurrence of evolution during his trip to the Galapagos Islands.

What he noticed was that while all of the islands had a type of finch bird, the different types scattered across the islands were not exactly the same. For example, a finch on an island might have a shorter beak than another finch from another island.

Darwin concluded that evolution was due to the species adapting to meet the particular conditions of a certain changed environment. For example, while the finches across the islands might once have been a single species from the mainland, due to their being in an alien environment they had to adapt and that's where we see evolution.

The traits that might have been useful on one island, like having a longer beak for eating nectar out of flowers, might not have been useful on an island bereft of flowers.

So, on the island where flowers would be plentiful, longer beaks had the advantage of survival and lasted longer than other organisms without longer beaks to pass on their long beak genes.

Over time, the birds on that island could grow to have on average, longer beaks than the birds of the mainland, simply because longer beaks meant a greater possibility of survival. And so there would be more long beak genes in the collective gene pool of the island's birds than those of the mainland.

This can also be seen in types of viruses. A virus

in its host's body may be eradicated through the use of antibiotics. However, some of the viruses might have a gene that helped them have a resistance to the antibody.

Then because of the use of the antibody, all of the viruses without the gene for resistance would die, while the virus that had the resistance gene would survive to pass on the gene to the next generation.

Then, like the birds earlier, over time the viruses would have developed a resistance to the antibiotic and the particular medicine would not work on it anymore.

As antibiotic resistance spreads, it makes the antibiotic less effective so new treatments have to be developed to combat the resistant bacteria. If there are no new drugs to fight the bacteria, treatment may be impossible. That's why it's important for individuals to use antibiotics only when needed, and not for a small cold.

Some viruses are already resistant to all of the antibiotics that we have discovered or made to this day. We call them superbugs, and they are the cause of 35,000 deaths a year in the US alone.

From these pieces of information, Zeus may have

been trying to spread his powerful, god of the sky genes, amongst mortals, so that in the case of the Titans rising and rebelling again, the mortals and gods could defeat them, by the survival of the fittest.

Zeus took in mind Darwin's theory and as the most powerful god, had a 'duty' to have as many powerful children as he could, lest an enemy of Olympus should arise. Some of these enemies were as powerful as the gods themselves.

One such example is the monster Typhon. Typhon was the son of Gaea and Tartarus, the pit from which all evil springs. As the son of such powerful Primordials, he also fathered many of the monsters that plague the ancient Greek myths, with his wife Echidna. Famous examples of his children include the Chimera, the Cerberus, and the Nemean Lion.

When Typhon rose, he was half as tall as the sky and his arms so long they stretched from the coldest north to the hottest south.

When the gods first saw Typhon, they were struck with fear and fled and hid in various places of the world in animal forms. Each god turned into their sacred animal. In fact, the Greeks claim that because some of the gods went to Egypt, that was where their animal-headed gods were born.

However, Zeus stood his ground. Although Zeus tried to fend off Typhon, his power was no match for the giant without the help of the other gods.

Typhon took Zeus's weapon, the thunderbolt, and to make sure the king of gods would not bother him again, removed his sinews.

With his sinews removed, Zeus was incapable of movement. But his son Hermes and his friend Aegipan came to Zeus's rescue. While the satyr Aegipan distracted Typhon with his beautiful music sprung from his pan pipes, Hermes stole Zeus's sinews along with his thunderbolts and gave them to Zeus.

With his health restored and his weapons reclaimed, Zeus beat Typhon and threw a mountain on top of him. This mountain was Mount Etna and legends say that when the mountain rumbles, Typhon is stirring in anger against Olympus.

Chapter Five :

Poseidon

With that he rammed the clouds together—
both hands clutching his trident—
churned the waves into chaos,
whipping all the gales from every quarter,
shrouding over in thunderheads the earth and sea at once—
and night swept down from the sky—
East and South Winds clashed and the raging West and North,
sprung from the heavens,
roiled heaving breakers up—

02

Poseidon, when rolling the die to decide realms with his brothers, got the second highest roll and chose the sea as his domain. His symbol was his weapon of war, his trident. Poseidon's sacred animal was the horse as he had made the animal himself.

One of Poseidon's various titles was the Earthshaker. He was called that because, with the help of his magical trident, he could shake the world because the Greeks thought that the world was encompassed by the seas.

His wife was Amphitrite who was a sea spirit called a Nereid. Poseidon decided to choose her as his wife among her 50 sisters. However, Amphitrite didn't want to marry Poseidon so he chased her around, trying to persuade her to marry him. Amphitrite couldn't stand it so she fled to the edge of the world near the Atlas Mountains where Atlas was being forced to hold up the sky.

The god of dolphins, Delphin, came to her and explained all of Poseidon's good attributes to persuade her to marry him. And persuade her he did. Amphitrite became the wife of Poseidon and the queen of the sea.

He fathered a son, Triton who was half human and half fish, like a merman, but had two tails instead of just one. Triton acted as his father's herald, blowing his conch horn and parting the waves. Poseidon also fathered two othedaughters named Rhode and Benthesicyme.

However, much like his brother, Zeus, Poseidon didn't stay faithful to his wife and sired many other children and monsters such as Theseus, various Cyclopes, and the famed winged horse, Pegasus.

Another famous myth about Poseidon was his fight with Athena for the Greek city of Athens. Poseidon and Athena both wanted to be the patron of Athens because that would result in more sacrifices for them.

The ancient Athenians suggested a contest, for which

of the two immortals gave them the better gift, in order to select the immortal with the better gift as their patron.

Poseidon summoned horses for his present while Athena gave them the olive tree. While horses were useful, they were not adapted to the rocky and dry Athenian climate and were of little use.

However, the olive tree could be grown with little effort and had many good uses such as olive oil which could be used for bathing and cooking. They could also make a profit selling the olives. The Athenians choose Athena for their patron and that was the beginning of Poseidon's rivalry with Athena.

However, to appease the two gods, the ancient Athenians, while building a temple for their patron goddess on the Acropolis, they also built a temple for Poseidon.

Like his brother Zeus, Poseidon took advantage of his shapeshifting abilities to have as many children as he could. These shapeshifting skills were very versatile, used in every type of situation.

The shapeshifting skills of the gods remind me of the skills of the viruses. In fact, even now, as I write this sentence, a new version of the Covid 19 virus is spreading, called the Delta variant.

Let's now look at some of the differences between various pathogens, and which ones of them have shapeshifting abilities to rival even that of the gods.

The first pathogen that we need to look at is bacteria. Bacteria are single-celled organisms that have a cell wall. They are usually between 0.5 nanometers and 0.5 millimeters in size. Also, because they have enzymes, bacteria can do their metabolism.

The genetic material of bacteria is located within the cytoplasm, a thick solution that fills each cell and is enclosed by the cell membrane.

Most bacteria have something called peptidoglycan. This is the material that makes up the cell walls of the bacteria. Then, some bacteria have a sticky film on their outer surface. This film helps them stick to teeth or skin belonging to other organisms so that they can infect them.

Bacteria exist almost everywhere in the world and they breed through asexual reproduction, more specifically, dichotomy. If their surroundings are a good breeding ground, they can breed very quickly. Under prime conditions, some species of bacteria can produce once every twenty minutes.

There are about ten thousand species of bacteria, but most of them are harmless to humans. However, some pose a very severe threat to humanity. For example, tuberculosis, tetanus, stomach ulcer, and bacterial food poisoning are some examples of diseases caused by bacteria.

The main weapon that we humans use to defeat bacteria is antibiotics. Antibiotics are originally substances formed by plants and animals to fend off bacteria competing with them for the same resources.

The second pathogen to look at is the virus. Viruses are different from bacteria in which they are so small we cannot see them with a light microscope. We need more powerful microscopes to do the job. The structure of viruses is a protein shell surrounding genetic information.

Unlike bacteria, viruses don't have any enzymes, so the viruses have to wait to infect a host cell first before they can metabolize. The virus hijacks the host cell and using its protein-making mechanism, forces it to make copies of the virus. This is the parasitic way of the reproduction of viruses. There also is a lot of mutating and change in the DNA of the virus when the host cell makes the copies of the virus.

When enough copies of the virus are made (about 50~100), the newly formed viruses break out of the cell and move on to infect other nearby healthy cells.

Some diseases caused by viruses include AIDS, the common cold, the flu, smallpox, SARS, MERS, measles, hepatitis, and more.

The SARS disease was prevalent from November 2002 until July 2003. It was spread all around the world by the RNA spread corona virus. People infected with the virus experienced a fever, muscle pain, and headaches, along with other symptoms.

The HIV virus was different from other viruses in the way that it took the helper T cell as its host. The role of the helper T cell will be explained in greater detail in the next chapter, but in short, the helper T cell detects which type of virus it is through the virus's antigens and activates the appropriate immune system.

However, because the virus takes place inside the helper T cells, the adaptive immune system cannot differentiate between different viruses and therefore cannot respond quickly and effectively.

The fact about AIDS is that the HIV virus itself does not kill its patients outright. It just lowers the ability of the immune system so that other viruses can take the weakened body out.

Diseases like these are cured with anti-viral drugs. However, anti-viral drugs have several side effects. Because viruses reside in the host cell, even if scientists develop an anti-viral drug, it has adverse effects on the host cell as well as the virus.

Also, because viruses have a high chance of mutating, even if there were an anti-viral drug, it would not be as effective or efficient in treating the disease.

The third pathogen is the protist. Protists are mostly made up of single-cell organisms and are spread by their animal hosts, called vectors. Some vectors include the mosquito, the flea, rats, and more.

Diseases spread by protists include malaria, sleeping

sickness, and amoebic dysentery.

The disease malaria infects the body by way of its host species, mosquitoes. The newly transmitted malaria goes inside the red blood cells. It takes the red blood cell as its host and after making enough copies of itself, it destroys the red blood cell. The poison that is released when the red blood cell is ruptured makes the body's temperature increase up to 40 degrees Celsius.

Another type of protist is fungi. Fungi are multicellular organisms and they reproduce with spores in moist conditions. The spores of fungi develop on the skin, or inside digestive systems, or respiratory systems.

Diseases caused by fungi include athlete's foot, candidiasis, and more. They can be infected by coming into contact with a foot or body part that has already contracted the disease.

The last pathogen that we need to look at is the prion. A prion is an incorrectly folded protein that can transfer its incorrect shape onto the same type of protein as itself.

If a normal protein comes into contact with a prion, it causes an ever-growing chain reaction, with more normal proteins receiving the incorrect shape.

Diseases caused by prions include mad cow disease and the Creutzfeldt-Jakob disease, which affects humans.

In the case of humans, if we consume an infected animal brain or infected animal tissue, we get infected with prions as well. Prions have a very stable structure, and will not be destroyed with either boiling or steaming food. The best way to prevent getting infected with prions is to avoid infected meat in the first place.

Out of the four types of pathogens that we looked at; the virus is the pathogen with the shapeshifting ability to rival the gods. The virus has a high potential of mutating and forming a new type of virus.

Chapter Six :

Hades

No one can hurry me down to Hades before my time,
but if a man's hour is come,
be he brave or be he coward,
there is no escape for him when he has once been born

02

Hades rolled the least dice number out of his brothers and so he was left with the Underworld. His symbol was his helm of darkness, which he received from the Cyclopes while his sacred animals were screech owls and black lambs. The Underworld was a place in Greek mythology where spirits went after they died.

There were three parts of the Underworld: Asphodel, the Fields of Punishment, and Elysium. There was another part of the Underworld which was apart from the rest: Tartarus. Tartarus was the gods' prison for their enemies seeing that it took nine days to fall from the mortal world to Tartarus, much like how it took nine days to fall from Olympus to the earth.

There were also five rivers that flowed into the Underworld. They were the Styx, the Acheron, the Cocytus, the Phlegethon, and the Lethe.

The Styx was the river of hate and the barrier between the mortal world and the Underworld. Because the Styx river goddess helped the gods in the war against the Titans, the gods rewarded her by making an oath on her waters unbreakable. It was the most solemn oath anyone could take.

Some legends say that the Styx also had the power to make one invincible. Achilles, the warrior son born of Thetis, was dipped into the Styx as an infant by his mother holding onto his ankle and was thus made invincible except for the spot where his mother held him.

The Acheron was the river of Pain and from it sprang the Styx and the Cocytus. It was the ultimate punishment for the spirits that deserved it and the spirits were put into the river and were swept deeper into the Underworld. It also had its starting point somewhere near Epirus before it flowed into the Underworld.

The Cocytus was the river of Wailing. Spirits who

wanted to stop the punishment were taken to the Cocytus once in a while and pleaded to those who they had wronged to forgive them. If they were forgiven, they were let out of the Fields of Punishment, and if not, they stayed.

The Phlegethon was the river of Fire. Spirits were swept into it and burned alive. It has an interesting property to heal because the tortured spirits needed to endure more torture and so the Phlegethon healed them to prepare for more excruciating torture.

Lastly, the river Lethe was the river of Forgetfulness. If a person were to drink from its waters, they would forget everything about themselves and their previous lives. Unlike the other legendary rivers of the Underworld, the Lethe was a kind of blessing to the spirits of the dead. Spirits gathered around the river to drink from it and forget their past life in order to be reborn in the mortal world again as a different person.

The Lethe was said to be curved around a cave where the god of sleep, Hypnos dwelt. Around the mouth of the cave were said to be poppies and other sleep-inducing plants.

When the spirits of the dead got to the Underworld, they were ferried across the River Styx, which served as a boundary between the mortal world and the Underworld, by a demon named Charon for the fee of one coin where a person's relatives would place a coin under their tongue as a funeral rite.

The spirits who did not have a coin were left to wander aimlessly. Some even went back to the mortal world because they did not accept that they were dead. When they were across the Styx, the spirits were judged by the three judges who looked through mortal lives to see if they did anything good or bad.

If a person lived an average life, they were sent to the Fields of Asphodel. If they did something good, the judges rewarded the spirit with Elysium where heroes and kind people were. If they did something bad, they were sent to the Fields of Punishment where they were tortured for eternity.

The main goal of Hades was to not let any spirits escape from the Underworld and back to the mortal world. To do this, he got the help of Cerebus, a big three-headed dog, and the Furies, who were born of Ouranos's blood when he was killed by Kronos.

Hades was also in charge of deciding the punishment for the wicked spirits of the dead. For example, there was a Greek king named Tantalus who was so mad the gods did not let him taste ambrosia, even when he was on Olympus, he invited them all to his palace and cooked them his son for dinner.

When the gods found out, they destroyed him and sent him to Hades for punishment. Tantalus's punishment from Hades was standing in a pool of water with a fruit tree hanging over his head. However, when he tried to reach for the food, the branches raised his target away from him and the water receded when he tried to touch it. His name is now used to mean temptation without satisfaction.

Another example was when a person named Sisyphus tried to cheat death. Sisyphus caught the god of death, Thanatos, off guard and hid him under his bed. However, the god of war Ares saved Thanatos and Sisyphus was sent to the Underworld.

Even in death, Sisyphus had not learned his lesson

and tried to cheat death once more by telling his wife to not give him a proper burial and pleading with the god Hades to let him go back and scold his wife. Hades agreed and Sisyphus went back to the mortal world. He got his body back and spent a couple of years avoiding death.

But he was eventually discovered of his treachery and was punished by rolling a boulder to the top of a hill. Sisyphus tried hard but every time, the boulder was pushed back near the top of the hill and left Sisyphus doomed to spend all of eternity rolling his boulder.

Unlike his brothers, Zeus and Poseidon, Hades remained mostly faithful to his wife, Persephone. However, he did have an incident

Hades was the personification of death, much like how his brother Poseidon was the personification of the sea. I personally find the idea of Hades and death comfortable for two reasons.

The first is that death is fair. Like in the myths of Tantalus or Sisyphus above, people get exactly what they deserve based on how they lived their lives. It also helps that the idea of there being a world after death helps people become more kind to other people because of the fear that they might be punished.

Then, after being pious for a long time, it becomes a habit and you do it unconsciously. Like Steve Jobs said, you make your habits, but in the end your habits make you. People pretend that they are good people and over time, they actually become what they pretend to be, or at least, that is what I want to think.

The second reason that I find death comforting is that death is cold and unforgiving. No matter what you do, you still find yourself crossing the river Styx and receiving judgment.

Everyone is destined for death and it is a fate that none of us can avoid despite our best efforts. The best we can do is to follow our hearts and intuition so that there will be no regret when each of our times come. You cannot prolong the day but you can enjoy it to your fullest.

There was a time before we mammals came to rule the world, such as the age of the dinosaurs. And although our death will mean the end of everything for ourselves, the universe, with its hundreds of thousands of galaxies and an even larger number of stars, will still go on.

We are but a collection of cells, fighting for our survival, exploiting each other for energy to live to see another day. While this fact may be unsavory to state, it is still the truth. It is the harsh nature of life.

Most of us humans are afraid of death, and we look for ways to extend our limited lifespan. Thanks to the development in the field of medicine and engineering, humans can now live up to a century or more.

However, it seems that our biological capabilities of longevity stop there. Or does it? Developments in nanotechnology could have nanobots flowing through our veins within the next few decades or so, helping our original immune system battle and fight off illness.

But how does our immune system function anyway? Our immune system is really made up of two systems called the innate immune system and the adaptive immune system.

The innate immune system, as can be inferred from its name, is the immune system that all people and almost all animals along with organisms are born with. In contrast, the adaptive immune system is the one that evolves and adapts during the course of a person's life.

An example of the innate immune system are surface barriers. In humans, our skin is our surface barrier. In insects, it is their exoskeleton, and in plants their waxy cuticle covering their leaves.

However, because we cannot be completely closed off from the rest of the world, due to nutritional and survival needs like energy, the openings in our bodies are well protected.

An example of this would be the hairs in our noses, along with other secretions like mucus that trap bits of harmful pathogens and dust from getting inside our bodies. Another would be the coughing and sneezing that our bodies do, due to several survival mechanisms at work, such as the unconditional reflex from the midbrain. Also, our tears help wash out pathogens along with the good bacteria in our intestines doing the same.

It is not only these physical barriers at work, however. There are also several chemical barriers as well. To illustrate, there is a covering of lysozyme on the surface of our skin.

The lysozyme helps break down the cell walls of bacteria and so helps stop bacterial infections efficiently. The chemical lysozyme is also in our saliva and it helps down the bacteria in our food. Also, the acid in our stomachs helps destroy the bacteria that the lysozyme in our spit did not get rid of.

One important part of the innate immune system is the inflammatory response. The inflammatory response is triggered when the skin or mucous membrane is harmed and pathogens come in through the injury.

In this case, let's make it a splinter that damaged the skin. The first signs of an inflammatory response would be the surrounding skin turning red and having a fever. The surrounding skin turning red is due to the mast cells widening the capillaries to allow more white blood cells to access the injury and take care of the pathogens.

The white blood cells would use a process called phagocytosis, a cellular process for ingesting and eliminating foreign particles. In this way, the white blood cells, with the help of the mast cells, destroy the pathogens.

Then, we go into the adaptive immune system, where things get a bit tricky. In the case where bacteria made it inside our bodies, white blood cells would be deployed first. White blood cells are still part of the innate immune system, not the adaptive immune system.

If the deployed white blood cells are overrun by the invading pathogens, macrophages are then sent out into the field. This is when the true adaptive immune system is activated.

Macrophages are specialized cells involved in the detection, phagocytosis, and destruction of bacteria. The additional function of macrophages is that they present antigens to T cells and release cytokines, small proteins that control the growth and activity of other immune system cells.

The adaptive immune system is divided into cell-mediated immunity and humoral immunity.

The difference between the two is that the cell-mediated immunity system has the T lymphocyte or T cell, and in humoral immunity, there is the B lymphocyte or B cell.

The T cell goes directly to the cells infected by the pathogen and then destroys it, while the B cell releases antibodies and memory cells into the bloodstream to wash out any pathogens in the blood.

The T cell is made in the marrow and is later moved to the thymus to reach maturity. On the other hand, the B cell is also produced in the marrow but unlike the T cell, also reaches maturity in the marrow.

For the immune system to determine whether the pathogen spreads by cell or in the bloodstream, the macrophages from earlier bring a small piece of the pathogen for the helper T cell.

The helper T cell then goes to the T cell if it is a cell that spreads by cells and goes to the B cell if it is a pathogen that spreads in the bloodstream.

The T cell, rather than moving around and destroying pathogens like the B cell, stays in place and multiplies. It is divided into killer T cells and memory T cells. The killer T cells produce antibodies that target the pathogen, while the memory T cells take note of the invading pathogen and remember it.

When the pathogen invades the body somewhere later in the future, the stored memory T cells are activated and they immediately transform into the killer T cells. This type of process is the fruit of the adaptive immune system. They recognize the pathogen and react to it faster.

In fact, this reaction time gets faster every time. This is the basis for vaccines. Vaccines have a small, harmless piece of a pathogen inside them and this triggers the adaptive immune system. In this way, even if the same pathogen invaded again for real, it would not be any danger because of the memory T cells already formed due to the vaccine.

Back to the original topic at hand, nanobots in our bloodstream could help both our innate immune systems and our adaptive immune systems by also targeting and destroying foreign pathogens.

But what if there was another way to extend our mortal lifespans? Even to the edge of infinity?

This technology is called mind uploading, and no, it doesn't exist yet so don't get your hopes up. However, the idea of mind uploading involves a process of scanning and imaging the brain, and uploading that copy to the digital world.

This would result in one's consciousness being immortalized in a digital form. This process would require the brain's synapses and neurons to be scanned and replicated in enough detail in order to replicate the person's unique brain.

The neurons are the basic cells that make the brain function by being designed to receive information from gland cells, decide what to do, and send that order to the other neurons and muscle cells.

The neurons are divided into the cell body, dendrites, and an axon. The cell body carries genetic information, does metabolism to make energy for the entire cell to function, and maintains the neuron's structure.

The dendrites are branch-like extensions at the front of the neuron in order to receive information from other neurons. The dendrites then pass the information to the soma, the longest part of the neuron and it then, in turn, passes along the information to the other neurons and muscle cells.

The neurons are also divided amongst themselves as the sensory neuron, the interneuron, and the motor neurons.

The sensory neuron receives information from the five senses, sight, smell, taste, touch, and sound, and passes the information along to the interneurons. The interneurons are responsible for deciding how to react to the information, and pass the decision along to the rest of the interneurons and finally to the final stage of neurons, motor neurons. The motor neurons pass the reached decision of how to deal with the stimulus from outside the body to the muscle cells, which then carry out the order.

In order to upload a digital copy of ourselves, we need to map out all of our 86 billion neurons and the individual connections between each and every one of them.

In order to achieve this goal, we need to advance two technologies. The first is how to image the brain and the second is to prepare the download space and processing capabilities of the 86 billion neurons and the connections between them.

In the area of imaging the brain, the most powerful, non-invasive technology available to image the brain is the MRI. And even the MRI can only see the brain up to half a millimeter at best.

For a computer to even see the neurons would require a powerful enough machine that could see up to the thousandth of a millimeter. With this, it is best to say that this sort of technology will only be available in the far future.

On the other hand, if we say that we already have the technologies to image the brain, the computer processing all of the information our brains hold into ones and zeros is a closer target.

However, there are still much more concerns than our limited technology. With the possibility of living forever, ethical questions inevitably arise. Questions such as how this technology could be abused, how should the eligible participants be selected, and much more.

Also, is it worth it for the copied identities to live in a virtual reality, albeit one that looks and feels just like the real world? Is the reality we see around us even real?

To borrow a page from Platonian idealism and Kantian philosophy into the mix, all that we perceive to be real is nothing but reacting to a very narrow scope of stimuli, such as sound waves, light, chemicals in both solid, liquid, and gas forms around us, which is then observed and picked up by our senses, and processed in the brain to make it seem like we perceive reality for what it is.

There is so much that we do not experience in the narrow scope of our senses. For example, while birds can sense the Earth's electromagnetic field for their navigation, we humans cannot imagine what the Earth's electromagnetic field would even feel like or how to use it.

We can only use our sensors to even pick up on the signals and stimuli that other animal races find natural to observe.

In many ways, it is like the difference between humans being able to distinguish between a cat and a dog all the time, while on the other hand, an AI system requires many samples of different forms of cats and dogs in order to even tell the difference between the two, and even then, it is not 100% correct.

On this note, I end this chapter of death and immortality, morality and unethicality, and knowledge and ignorance.

Chapter Seven :

Demeter

Queen of fragrant Eleusis,
Giver of earth's good gifts,
Give me your grace, O Demeter.
You, too, Persephone, fairest,
Maiden all lovely, I offer
Song for your favor.

Demeter was the second-born child of Kronos and Rhea. She was the goddess of agriculture. Her sacred animal was the serpent, and her symbol was the sickle. The sickle was said to be molded from the weapon of her father Kronos, the scythe he used to dethrone Ouranos.

She had various children with different gods. Zeus wanted to marry her but she refused and turned into a snake to hide from him. However, he turned into a snake himself and chased after her. From the union of Zeus and Demeter, the goddess of spring, Persephone was born.

Another example was when Demeter turned into a horse to escape from Poseidon who also wanted to marry her but got caught. She gave birth to the fastest horse, Arion who was immortal and could run over both water and land and also was blessed with the power of speech, and his twin sister, Desponia, who was made a priestess of Demeter.

The god of the underworld, Hades wanted to marry Persephone born from Demeter and Zeus. When he went to Zeus for his blessing, Zeus could not help him because they needed the consent of Demeter as well.

However, Zeus decided to help Hades abduct Persephone. Zeus found Persephone picking flowers in a meadow and upon seeing her, he made another field of flowers right next to her filled with the most beautiful flowers she had seen. She sneaked away from her nymph companions who were sent by Demeter to protect Persephone.

When she was in the field of flowers, Zeus made another field filled with flowers which made the current flowers of the field, where Persephone was in, pale in comparison. It went on and on like this until Zeus had led Persephone to the place where Hades would kidnap her.

Hades rose from the Underworld in his chariot and abducted her and brought her to his palace in the

Underworld. Persephone cried and screamed for her mother Demeter but she was trapped.

Meanwhile, back on earth, Demeter was searching frantically for Persephone. She punished the nymphs who were supposed to take care of Persephone by punishing the water nymphs by cursing them to become sirens, water demons who were supposed to sing to sailors and lure them to their deaths.

When Demeter came across the place where Persephone was kidnapped, the goddess of magic, Hecate came out to help her. Hecate used her torches to light up the sky and help Demeter. But, for all their efforts, they could not find anything.

During her search for Persephone, Demeter had many adventures. One of her adventures was with the kingdom of Eleusis where she disguised herself as an old woman and entered the kingdom where King Celeus and Queen Metaneira reigned.

The king and queen were kind to Demeter even though they did not know she was a goddess because, in ancient times, the gods would go down to mortals disguised as travelers or beggars, and as a Greek citizen, it was their duty to offer them a place to sleep and a meal to eat.

The king and queen had two sons, one was Triptolemus, who was already of age, and the other was Demophon, still an infant. Triptolemus offered to send riders in the four directions of the earth to find out what had happened to Persephone.

Demeter was taken aback at the kingdom's kindness but recovered quickly. She tried to make the infant prince of Eleusis, Demophon, immortal to repay the kindness given to her.

In order to make Demophon immortal, Demeter fed the baby ambrosia and nectar, the food of the gods eaten on Olympus. She also put powerful protective spells on him and placed him in the fireplace. The fire would slowly burn away Demophon's mortality, and, combined with the ambrosia and nectar, would make him immortal.

However, on the third and final day of Demophon's transformation, Queen Metaneira intervened by checking on her baby, and not knowing anything about Demeter's plan, snatched the baby out of the fire.

That night, Triptolemus returned with news about Persephone from a god named Helios. Demeter was not able to finish the ritual but later made the other prince, Triptolemus, Demophon's older brother, and the son of Celeus and Metaneira, the god of farming instead.

Demeter, according to Triptolemus's information, went to visit the god of the sun, Helios because from his sun chariot in the sky, Helios could see everything that was happening. Helios was a Titan but stayed neutral during the war between the Gods and the Titans along with his sister, Selene, god of the moon. Helios revealed to Demeter that her daughter Persephone had been kidnapped by the god of the Underworld, Hades.

Demeter went to Zeus but he said that Persephone was already married to Hades and that she could not get her back. Demeter made all the crops and plants wither and sent the entire world into famine. Pressured by all the mortal cries in agony from starvation from the earth, Zeus gave in to Demeter and ordered Hermes, the messenger god, to go to the Underworld and fetch Persephone back.

But it turned out that Hades had tricked Persephone

into eating a third of a pomegranate during her stay in the Underworld. It was a rule of the Fates that if a person eats food in the Underworld, they are destined to stay there forever. Demeter and Hades had a compromise through Zeus.

Hades would get Persephone for one-third of the year for the third of the pomegranate that she ate, for the rest of the year, she would stay with Demeter. During the months that Persephone was away from her, Demeter would grow cold and unhappy again and forget to cultivate and fertilize the earth once more, which would essentially be winter. So, the coming and going of Persephone created the seasons.

Most people, when they read the story of Demeter and Persephone, think that Demeter went a little overboard with starving the world but starving the world was the only leverage Demeter had over Zeus and the rest of the gods. It was an advantage that she had to use, lest she lose her daughter forever.

Another myth involving Demeter, less known than the myth of Persephone, was the myth of Erisikhthon.

Erisikhthon was a prince who decided that he wanted a new palace for himself. He went to the sacred woods of Demeter where the tallest trees were and tree spirits called nymphs celebrated in honor of Demeter.

Erisikhthon and his servants went to cut down the trees when the goddess herself appeared in front of him in the form of a young girl who tried to stop them by telling them that the woods were sacred to Demeter and they would be punished for their crimes against the goddess and her fellowship of nymphs.

Erisikhthon pushed the young girl aside and went on his path with his servants when the goddess revealed her true form to him and cursed him. Erisikhthon went running home crying and begging for forgiveness. However, it was too late.

When Erisikhthon got home, he noticed that he was hungry. He ate a feast all by himself but when he was done, he was still hungry. So, he ate and ate and ate but his hunger was not satisfied.

His money started to run out when he ordered servants to fetch him more food. Erisikhthon's hunger was unquenchable. He was so desperate for food that he would gladly resort to anything to get money which meant more food. Erisikhthon even sold his own daughter as a slave. However, Demeter took pity on the girl and sent her to Poseidon. Poseidon took her under his wing and made her a servant in his underwater palace.

Erisikhthon was left poor and friendless. He went back to Demeter's grove thinking that he could get the goddess to remove the curse. But Demeter ignored him. In the end, he started to gnaw on his foot and arms, and legs. When the sun rose again, he was gone.

All of the gods have at least one myth that shows them as stern rulers of the world. These kinds of awe-inspiring, fear-inducing myths made sure that the ancient Greeks honored the gods and did everything in their power to not anger the gods, for they believed that they were a force not to be reckoned with.

Demeter punishing Erisikhthon was just because it was his fault. She was ready to forgive the incident in the beginning by turning into a young girl and warning him. However, it was his fault that he decided to ignore her subtle warning or advice and therefore seal his fate.

While the ancient Greeks believed that the coming and going of Persephone caused the seasons to change, due to developments in science, we know that the changing of seasons is because of the Earth's yearly revolution around the sun.

What causes the seasons? While some people feel that summer is due to the Earth being closer to the sun while in winter the Earth is further away, this is not true. Because the Earth has a nearly circular orbit, Earth is the same distance from the sun at all times.

The real reason for the seasons is due to the Earth's tilted axis. The Earth is tilted at an angle of about 23.5 degrees. While this difference in angle is not much compared to the sun, which is 150.55 million kilometers away, it is everything.

In the perspective of the northern hemisphere, when the sun is tilted towards the Earth, it is summer. When the sun is tilted away from the Earth, it is winter.

Now the reason why I added 'in the perspective of the northern hemisphere' is because the two hemispheres have the opposite season at the same time. When the sun is tilted towards the northern hemisphere, it is tilted away from the southern hemisphere, which causes summer in the north and winter in the south.

This is why in Australia, people celebrate Christmas with sand snowmen, while we in the north celebrate with our snowy counterparts.

If the Earth's axis did not have a tilt at all, reality would be very different from the one that we have now. For one, the sun at noon would always be directly above the equator, and day and night would be the same length everywhere on Earth. There would be no daylight saving time either.

Also, there would be no seasons. The equator region would always be hot, while the North and South poles would be eternally cold. The further away from the equator a traveler went, the cooler he would be while the closer he got would result in an increase of temperature.

Hence, the tilt of the Earth's axis explains why we have seasons, along with why days are longer in summer than in winter. However, we still need to ask the question of why summer feels so hot while winter feels so cold.

The answer to this new question relies on the angle of the sun's rays. In summer, the sun is high up in the sky, and the angle that its rays beam down on the Earth is big.

On the other hand, in winter, the sun is relatively low in the sky owing to the Earth's tilt and therefore light reaches Earth at a shallower, smaller angle. This means that the sun's rays are spread out over an even larger area, and so the Earth gets a smaller portion of the rays than it received back when it was summer.

Chapter Eight :

Hera

Golden-throned Hera, among immortals the queen,
Chief among them in beauty, the glorious lady
All the blessed in high Olympus revere,
Honor even as Zeus the lord of the thunder.

Hera was the last daughter of the Titans Kronos and Rhea, born between her sister Demeter and her brother Hades. She was the goddess of marriage and the queen of the gods and Olympus.

Hera was married to Zeus but did not initially plan on marrying him. Zeus had to trick her. He disguised himself as a cuckoo bird and summoned a storm in order to get sympathy from Hera. Hera was persuaded to marry him because of his ingenuity, and hence, her sacred bird became the cuckoo bird.

Hera and Zeus's marriage was one of the greatest events to happen in the ancient world. The heavens, earth, and seas cried out in joy, and every animal, regardless of if they were of the land, the sea, or the heavens, were invited.

During the wedding, the most magnificent wedding gift was a gift to Hera from Gaea, the tree of immortality. The tree was placed in the Garden of the Hesperides, far away to the west. The Hesperides, nymphs of the west and daughters of the evening, guarded the tree with a dragon called Ladon that Hera summoned.

Hera and Zeus had three children: Ares, the god of war, Eileithyia, the goddess of childbirth, and Hebe, the goddess of youth who served ambrosia and nectar, the food of the gods on Olympus.

Despite being the goddess of marriage and family, Hera is not known for her good deeds, but rather, chasing after Zeus and the women he chased after to get revenge. Since she was the goddess of marriage, she could not stand Zeus's cheating on her.

Though Hera stayed faithful to Zeus, Zeus often slept with other women, which for a goddess of marriage, was a big insult. So, she bided her time until Zeus put harsh punishments on the other gods, and succeeded in luring Poseidon, Athena, and Apollo to rebel with her against

Zeus.

Hera drugged Zeus and the other gods bound him to his bed while stealing his thunderbolt. However, Briareus, the Hundred Handed One who was rescued by Zeus, overheard the other gods planning to overthrow Zeus and used his many hands to untie him.

When Zeus woke up, he was furious. He punished the gods who had been part of Hera's plot to seize Zeus.

Poseidon and Apollo were sent to the city of Troy to become servants for the king. The king made them reinforce the city walls and fortresses, making the walls almost impenetrable.

However, Hera got the worst punishment of all. She was dangled over the mouth of Chaos, which was a pit opening to Chaos. Anything that touched it would be reverted back into their original forms, meaning nothing.

Zeus punished her by threatening to unhook the chains from Hera and letting her fall into the abyss. Hera would eventually be rescued by her son, Hephaestus. Zeus let her go only if she swore that she would never act against him again.

Another myth about Hera was the myth of Hera and Ixion. Ixion was the first mortal who made the first weapon. The gods were so impressed with him that they invited Ixion to dine with them on Olympus.

Ixion fell in love with Hera and he tried to rape her. Hera reported his actions to her husband Zeus, and to see if her report was true, he fashioned a cloud dummy that looked exactly like Hera and laid it by Ixion's side.

In the following morning, Ixion bragged to the other gods that he had slept with Hera. As his punishment, Zeus tied Ixion to a wheel, set it on fire, and tossed the wheel through the heavens where he would remain spinning and on fire for eternity.

Although Hera was the goddess of family, myths about her show, not her benevolence but her jealousy and vengeance on the women who had dared to sleep with her husband, much like how most myths about Zeus show his various conquests of women.

This example of the goddess Hera goes to show that none of us, not even the proclaimed "perfect, deathless gods who live on snowy Olympus" are perfect. However, it is debatable if her vengeance on Zeus's lovers is justified.

Although Hera had every right to be angry both as Zeus's wife and as the goddess of marriage, taking out all of her anger on the women that Zeus slept with is a bit too harsh.

If she had to be angry with someone, she had to be angry with Zeus, not the women, because, most of the time, Zeus forced himself onto them and they had no good choice to make unless they refused Zeus and were destroyed by him. However, since Zeus was the most powerful god on Olympus, she could not take her anger out on him.

If this scenario had happened in society today, Hera could have divorced Zeus instead of being stuck in a marriage that did not respect her rights.

As said in the Iliad, as the king of the gods, Zeus is more powerful than all the other Olympians put together. Therefore, if I was Hera and initiating the rebellion, I would have gathered more of the nature spirits like the nymphs and naiads.

The other non-Olympian gods are not to be ignored, however. For example, Morpheus, the god of sleep once put

Zeus to sleep as ordered by Hera to change the tide of the Trojan War.

What interested me here in the queen of Olympus's stories was how there was a cloud made to be shaped exactly like her.

In order to sate our curiosity, we must first know how normal clouds are formed. Clouds form in four main different ways.

The first is surface heating. Surface heating is when the Sun heats the ground, which in turn heats the surrounding air. Because the warmer air is lighter and less dense than the air surrounding it, it rises. Then it forms into a cloud because of expansion, as seen in the later paragraphs.

The second phenomenon of air rising is due to lower pressure. Lower pressure is when the air pressure of a region is lower than those of the regions around it. This difference in pressure causes the wind to flow from the higher-pressure regions to the lower ones.

When there is a lot of air coming in from all sorts of directions to a low-pressure area, the influx of air is forced to rise, because there is so much air from the high-pressure regions. This is known as the convergence of air.

Wind always flows from higher pressure to lower pressure, with no exceptions. Another interesting fact is that in the northern hemisphere, air comes into the lower pressure regions with a counter-clockwise direction, while air flows out of regions with high pressure with a clockwise direction.

The third reason for the rising of air is topography. When there are mountains, the air is forced to flow with the upward curve of the mountain and therefore goes up.

This is the reason why there is so much precipitation on one side of a mountain while very little on the other. The mass of air happens to go through expansion before it can completely go over the mountain.

The fourth and final reason for air rising is due to weather fronts. There are many types of weather fronts, such as cold fronts, warm fronts, stationary fronts, and more, but the one responsible is the warm front.

A warm front is formed when a mass of warm air pushes into a mass of cooler air. Because of the temperature difference between the two masses of air, the warm mass goes up, goes through expansion, and becomes a cloud.

What is interesting to note is that warm fronts and cold fronts form different types of clouds.

A cold front occurs when a cold mass of air pushes into a warmer mass of air. Due to the cold mass of air having a higher density than the warm mass, the cold mass pushes the warm mass faster, as opposed to the warm front where the warm mass pushes the cold one.

Due to cold fronts being faster than warm fronts, they push air upward very fast, which produces big looming cumulus clouds. On the other hand, with warm fronts, they push air slower and create wispy cirrus clouds.

When clouds are formed, it is because some air rises due to the four types of phenomena, as seen earlier above. Then, because the surroundings high up in the sky are low in both temperature and pressure, it goes through a process known as expansion.

Expansion is when the surrounding pressure drops quickly and so in order to maintain the pressure constant with the surroundings, the mass of air expands. This process can be explained based on the experiments of the scientist, Robert Boyle, and his Boyle's law.

Boyle's law states that when the surrounding temperature remains at a constant, the product of the multiplication between the current pressure and the current volume of a gas remains the same.

This Boyle's law is applied when clouds are made. The mass of air rises and, because of the lower pressure in higher altitudes, its volume starts to get bigger. Also, because the earth's radiant energy diminishes, the mass of air reaches its dew point when it gets higher.

When the air reaches its dew point, the moisture in the air condenses and becomes extremely small water droplets. These water droplets form together, and voila! You get a cloud.

And we move on back to our earlier subject: How was the cloud able to be a lookalike to the goddess herself?

The shape of clouds is more or less dependent on the wind. So, we can guess that Zeus made a cloud, with the four above phenomena, and used tiny gusts of wind to form it into Hera's exact shape.

Chapter Nine :

Hestia

Hestia, in all dwelling of men and immortals
Yours is the highest honor, the sweet wine offered
First and last at the feast, poured out to you duly.
Never without you can gods or mortals hold banquet.

02

Hestia was the firstborn of Titans Kronos and Rhea. However, the gods went by the birth order they were disgorged from Kronos's stomach and so Hestia was counted as the youngest sibling in most myths.

She was the goddess of the family hearth and presided over all sacrifices. As part of the goddess of the family hearth, she was also in charge of families' internal unity. To the ancient Greeks, Hestia represented anything that was domestic.

Hestia is not introduced in most myths as she was not the most glamorous and active amongst the gods. However, she was one of the most common. Because she was the goddess of the hearth, when the Greeks offered sacrifices to the gods, Hestia was always included.

Hestia was also responsible for keeping the Olympian family together as the goddess of family. She even gave up her throne on Olympus to avoid conflict of whether Dionysus should become an Olympian god or not. Needless to say, she had no enemies on the Olympian counsel and all of the other gods liked her.

One myth about Hestia was that both Poseidon and Apollo courted her for marriage but she rejected them both and swore herself to eternal virginity and became a virgin goddess, along with Athena and Artemis. To help the Greek family households, she needed to be pure. Zeus, as the king of the gods, accepted her decision and thus all the gods honored her wishes. In many ways, she was the opposite of her fellow Olympian, Aphrodite, the goddess of love.

One myth about Hestia was the myth about her and Priapus, the god of abundance at an Olympian party. When everyone was passed out and drunk from too much partying, Hestia went to the forest and passed out on the floor, the party being too much for her. Priapus, the god of abundance saw her and tried to sleep with her. However, a donkey one of the satyrs rode to the party brayed and woke Hestia from her sleep. She screamed and all the other gods came from the party to help her.

Hestia was a very peaceful and domestic goddess and was one of the few level-headed gods on Olympus. However, she was a little too agreeing and that made her lose her throne due to her not wanting the other gods to fight.

Hestia giving up her throne to Dionysus reflected a great change in the mortal world as well. When Hestia had her throne on Olympus, the Olympians were equal: six male gods and six female ones. However, when Dionysus came, there were now seven male gods and five female goddesses, making the female side unequal as opposed to the male side which grew more powerful. As with the dice roll before, this represents the oppression and inequality women faced in the time of the ancient Greeks.

Hestia also had a role in helping humans get fire. When Prometheus wanted to give his creations(humans) fire so they could live more comfortably, Hestia was all on board with the idea.

Since she was the official goddess of the hearth, she was in charge of tending to the hearth on Olympus. Back then, only the gods had access to fire. However, when Prometheus came to steal a few coals for himself to give to the humans, Hestia turned a blind eye to his actions and let him get away with it and give humans the gift of fire. Since then, Hestia was in charge of not only the hearth up on Olympus, but also the many hearths created on the earth.

The chapter on the goddess Hestia included how she helped humans. This chapter led me to think a little more about the nature of humans and inequality, the inherent function of our society.

The famous book, Guns, Germs, and Steel by Jared Diamond talks about how humans have evolved over the past years and how the distribution of certain plants and animals across the world helped create unequal differences in the world.

While at first, because there were only hunter-gatherer societies in the world, all humans were on some equal footing.

However, the start of inequality begins with the rise

of plants. Only a few plants were able to be easily farmable, with such examples including barley, rice, corn, and maize. The hunter-gatherer societies that lived in places where such crops were abundant were at an advantage to the societies that did not, all because they had a constant, reliable source of food and were able to stay in one place for a longer time.

The added extension of time made it possible for early cities to form, and as the population of these places kept growing, their overall power grew and they were able to start looting, pillaging other societies.

The other thing to take into consideration is animals. Unlike their plant counterparts, while the plants only varied in the difficulty of farming, only a few animals could be domesticated. Some examples included the dog, pig, cow, llama, horse, and chicken.

The cities that had a wide distribution of these animals had an advantage. Mainly because they were another reliable source of food and also because the people lived in close proximity with the animals, they developed natural immunities against the diseases that the livestock carried.

This helped because when they invaded regions where the animals were not as prevalent, the diseases helped them softly invade others. After all, the disease sapped the opponents of their strength while they were fine.

This led to Europeans quickly colonizing both North America and South America. Because the natives had no livestock and therefore no immunities to livestock-borne diseases such as smallpox, they succumbed in the face of the diseases the colonizers had inadvertently brought with them.

Because of these reasons, colonizers were able to assert themselves as more powerful than the natives who were invaded and that was when our problems of inequality started, such as slavery for example.

We can only hope that in the global society that we live in today, we would be able to put aside our past differences and be able to stand on equal footing with each and every person on this planet.

Chapter Ten :

Athena

Pallas Athena, the glorious goddess,
bright-eyed, inventive, unbending of heart,
pure virgin, saviour of cities, courageous, Tritogeneia.
Wise Zeus himself bore her from his awful head,
arrayed in warlike arms of flashing gold,
and awe seized all the gods as they gazed.
But Athena sprang quickly from the immortal head
and stood before Zeus who holds the aegis,
shaking a sharp spear:
great Olympus began to reel horribly
at the might of the bright-eyed goddess,
and earth round about cried fearfully,
and the sea was moved and tossed with dark waves,
while foam burst forth suddenly:
the bright Son of Hyperion stopped his swift-footed horses a long while,
until the maiden Pallas Athena had stripped
the heavenly armour from her immortal shoulders.

Athena was the goddess of wisdom, warfare, and crafts. Her mother was the Titan Metis, the Titan of Thought, and her father was Zeus. In contrast with her brother Ares who represented the bloodlust and the untamed parts of war, Athena represented the more tactical and logical sides of war.

There was a prophecy that the first child Metis would give birth to fathered by Zeus would be a girl and she would help Zeus enormously. The second child, however, would be a boy and have the power to overthrow Olympus, just like how Zeus overthrew Mount Othrys.

Zeus was very panicked that the possibility of his child overthrowing him as he had done to his father Kronos would come to pass. In his panic, Zeus devised a plan. He challenged Metis to a shrinking contest to see who had the better shape shifting ability. When Metis turned into a miniature version of herself, Zeus swallowed her and the baby she was pregnant with, whole.

However, much like how Kronos's children survived inside of him, Metis did as well. She stayed in her miniature version of herself and gave birth to the baby, Athena. When Athena was of age, she traveled to Zeus's head from his stomach and with the war weapons and armor her mother made her, made a loud racket.

Zeus had almost forgotten about Metis and thought it was clever of himself to trick her. He thought the whole thing was behind him and the possibility of a child greater than him erased.

However, one day, Zeus had a splitting headache and he asked Hephaestus, the god of metalworking, to split his head open. Hephaestus took an awl and split Zeus's head. Out came Athena, dressed in Greek armor and holding a shield and spear. Henceforth, she became the goddess of wisdom. Athena was also a virgin goddess, along with

Hestia and Artemis.

Some people called the wisdom goddess, Athena, while others called her Pallas Athena. The myth about the origins of the name Pallas Athena was when Athena had gone to the nymphs of Lake Tritones.

There, she made a friend who was almost as good at sparring as her. During one of their sparring matches when Pallas had the upper hand, Zeus looked at them and could not stand to see his daughter lose and appeared behind Athena with the Aegis that Pallas had shown interest in before.

Pallas was transfixed in awe and Athena, expecting Pallas to dodge, accidentally impaled her. In honor of her dead friend, Athena added her name to hers, making her Pallas Athena.

Another well-known myth about the goddess is how she made the Gorgon Medusa. Poseidon, who was defeated by Athena in the fight for Athens, was still sore over the wound and decided to get revenge by sneaking into Athena's sacred temple while taking a mortal girl, Medusa, with him.

While Athena could not get revenge on Poseidon

because he was a god, she could get revenge on Medusa. Athena cursed her to grow brass wings and talons, and worst of all, she transformed her face to one with boar tusks and her hair to snakes. Athena made Medusa so ugly that mortals would turn into stone if they looked her in the eye.

Athena would later get the hero Perseus to behead Medusa by assisting him with the killing with a mirror to see Medusa but to avoid getting turned to stone. Athena put the visage of Medusa on her shield so her enemies would cower in fright in battle.

The most famous myth about Athena would be how she made the first spider. In the kingdom of Lydia, a girl named Arachne was famous because of her talent at weaving. Arachne was bitter because the townspeople said that Athena herself must have taught her weaving but she believed that she had been born special with the gift. One day it was so much that she burst out, challenging the goddess to a weaving contest.

Much like Demeter who warned before her punishment, Athena did the same. She appeared as a wise old woman wanting to give Arachne advice. However, when she was brushed off and insulted once more, Athena accepted her challenge and appeared to Arachne.

When the two started their weaving contest, Athena wove a tapestry that depicted the Olympian gods sitting on Mount Olympus. Athena showed the gods as powerful, omniscient beings who deserved to be worshiped and treated with respect. She also wove in little warnings in the tapestry such as Salmoneus and Ixion being punished by Zeus.

On other hand, Arachne wove a tapestry that depicted the Olympian gods as foolish entities who did not know better than to chase mortal women and fight with each other by weaving a picture of Zeus kidnapping the princess Europa.

Athena had sworn of the river Styx to have a fair judgment of the two tapestries. When they were both done, Athena had to admit that they were woven with equal skill. However, Arachne was punished because of her disrespect towards the gods. Athena turned her into a spider to punish her for daring to believe she could be better than the gods and turned her into a spider to make sure she never touched a loom ever again.

This chapter on Athena, the wisdom goddess, prodded me to think about the goddess's domain: wisdom. There is so much that we do not know, but even the knowledge that we have collected from the start of the universe is commendable.

In one of Steven Pinker's (a person who I happen to admire very much) books, titled "Enlightenment Now," he talks about what we would be most proud of when we would enter into an intergalactic ego competition.

In the book, he says that while abolishing slavery is good, it was an evil that we had done to ourselves. As for the arts, the standards for appreciating music and art might be different from the other alien races.

So, the best thing that we could provide is science, or how much of the universe's secrets we unearthed. As beings existing in the same universe, every race could have a measure of how far, or maybe how not far our race has come.

Athena's punishments on Arachne were very reasonable. While I can understand things from Arachne's perspective, I believe that Athena was in the right. Although Arachne was punished for her pride, she was also punished for her disrespect against the gods represented in her tapestry.

This would hold true for us humans as well. If someone had insulted or wronged a member of our family, we would not hesitate to defend them.

In the current world, the meaning of family is ever-expanding. As seen in Jeremy Rifkin's book, Biosphere Politics, the author states that from the start of human history, we have been on a journey that has taken us into ever more inclusive domains of empathic engagement and collective stewardship.

As humankind progressed further into the future, we have gone from including our townspeople as family to, because of the two World Wars, regarding our own country as our family (what we know as nationalism), and finally, through the invention of the internet, come to thinking of the whole of humankind as our extended family.

However, we should not let the progress of empathy stop here. We should let our circle of family include the whole of the Earth, such as the plants and animals on our beautiful planet.

What we could achieve if we accomplished this is immeasurable. Most people would come to love and care for our planet. Even now, with the spread of Covid 19, the waters of Venice are already clear because human activity has ceased. Even dolphins are spotted in the canals.

If humans could do this kind of good in such a short amount of time, imagining what would happen if we managed to reduce our carbon footprint on the Earth is very difficult because the improvement would be so great. Humans could, after all these years of polluting, finally live in harmony with nature.

Another thing that the myths kindled curiosity in me was Medusa's stone gaze. Theoretically, there is another scientific explanation for her transforming gaze. It is related to the Laws of Thermodynamics.

While there are technically only three Laws of Thermodynamics officially, there is also a law number zero in existence, along with the original three.

The zeroth Law of Thermodynamics is concerned

with the act of heat transferring from the object with more heat energy to ones with lower amounts of heat. After a certain amount of time has passed, one would find that the temperature of the two objects has become the same value.

To expand further on the topic, there are three ways in order for heat energy to transfer from one object to another. The first is conduction. Conduction requires the two objects transferring and receiving heat to be in physical contact with each other.

Take the example of a stove and a pot. The heat would be made by the fuel of the flames (something like methane) and transferred to the pot by physical contact. The heat would travel along every molecule of the pot, transferring heat from the heat source to each other.

The second method of heat transferring is convection. Convection is not the two objects giving and taking heat by being in physical contact with one another but instead requires a medium for the heat to travel in.

There are examples of the occurrence beside the sea. At day, the air on top of the land is hot because the land has a lower heat capacity than the ocean, which has a higher heat capacity mainly because it is made up of water. And so, the air above the ocean is cool.

Because of this temperature difference between the two close bodies of air, convection occurs. The hotter air from land rises because warmer gases are lighter than their heavy counterparts.

The cooler air from the sea comes in to fill the space the hotter air-land left behind. This causes the wind to be in the direction of the sea to the land.

On the other hand, it is the opposite at night. Because the land has a lower heat capacity than the water, it cools down faster. Then, the roles from the day are reversed. The land is cooler than the sea, which, having a higher heat capacity than land, is comparatively warmer.

The warmer air from the sea rises while the cooler air from land swoops in to take its place. This causes the wind to be formed in the opposite direction than it did in day: from the land to the sea.

Then, the first law of Thermodynamics states that if heat is recognized as a form of energy, the total energy of the system and its surroundings are conserved. In other words, the total energy of the universe remains constant.

The first law of Thermodynamics is just a spin on the law of the conservation of energy.

An example of this first law of Thermodynamics is when you think of a bucketful of legos as the total energy of the universe. If a hydraulic press, (like in so many currently trending tick tok videos), took the legos and crushed them into tiny little pieces, the legos should, in theory, be able to be restored into their former glory with a bit of glue (while you would need a lot of glue to do so).

As the example from the legos, the total energy of the universe is a constant. While the total energy of a system and its surroundings may change according to the circumstances, the legos will always be able to be put back together. The total amount of energy remains the same.

Then we move on to the second law of Thermodynamics. Simply put, the second law states that if one has a closed system, any natural process occurring in the system always happens in the direction of the entropy in that system increasing.

In fact, all processes go in the direction of increasing

entropy. Entropy is interpreted as the disorder and chaos of a system.

Even in the process of salt water evaporating in order to make subsequent crystals, while the salt crystals formed are much more orderly than the salt ions in solution, the evaporated water is much more disorderly than in its liquid state. So, the total amount of disorder in the universe increases.

And finally, we must look at the third Law of Thermodynamics. The third law states that when an object approaches absolute zero, the entropy of that object becomes constant.

Absolute zero is theoretically the coldest possible temperature in the universe. It is so cold that not even the slightest tremor of the solids would exist. All objects would lose their energy and so their entropy would remain constant.

Absolute zero is 0 Kelvin, a form of measuring temperature where it uses absolute zero as its null point. It has the same magnitude as Celsius however, an increase of one Kelvin would result in an increase of one degree

Celsius.

Zero Kelvin is equal to -273 degrees Celsius and then equal to approximately -460 degrees Fahrenheit. Fahrenheit, unlike Kelvin and Celsius, uses a magnitude of 180 in contrast to their 100.

To put things into perspective, while water freezes at 0 degrees Celsius and reaches its boiling point at 100 degrees Celsius with a gap of 100, with Fahrenheit, things are a little different. Water freezes at 32 degrees Fahrenheit and boils at 212 degrees Fahrenheit, a difference of 180 between the freezing point of water and the boiling point of water.

Because of the concept of absolute zero, where all movement stops, I thought that maybe Medusa's stone transforming gaze had the same effect, bestowing the effects of absolute zero on her victims and making them seem like stone and lifeless. The cold would also help with preserving the bodies of the victims and furthering the illusion of stone.

However, because there were no records of the victims' bodies giving off an unbearable, burning cold, and because the victims turned gray and colorless, with no further information, we can conclude that Medusa's frightening gaze was not in fact, a heat siphoning machine, but a curse from the gods.

Chapter Eleven :

Apollo

O Phoebus, from your throne of truth,
From your dwelling-place at the heart of the world,
You speak to men.
By Zeus's decree no lie comes there,
No shadow to darken the word of truth.
Zeus sealed by an everlasting right
Apollo's honour, that all may trust
With unshaken faith when he speaks.

Apollo was the god of healing, medicine, archery, truth and much more. The ancient Greeks went undecided on Apollo or Helios being the sun god. In some myths the driver of the sun chariot is Helios, in others, it is Apollo.

However, because much of the myths surrounding Apollo has his role as the sun god, we will assume that while Helios was the original sun god in earlier times (like when Demeter was searching for her daughter Persephone), he was replaced with Apollo later on.

Apollo's mother was the Titaness Leto and his father was Zeus. He also had a twin sister, Artemis, goddess of the moon, who will be covered in the next chapter.

While Hera was relentless in her punishing of Zeus's

illegitimate children and their mothers, she was harshest on Leto. She sent a number of obstacles in the pregnant goddess's path, an example being the snake Python. Also, Hera put a curse on Leto, making it so that she could not give birth on the mainland or any of the islands.

However, there was one land that did not quite fit in Hera's curse: Delos. Delos was a floating island, having no connection to the bottom of the sea like other islands. As a result, it frequently drifted around, riding the waves.

Leto, after a long search, found Delos which granted her refuge and allowed her to give birth on the island. After the twins, Apollo and Artemis were born, Delos tied itself to the earth and became their sacred land.

After Apollo was born, he sought to seek revenge on the ones who had made his mother Leto's life miserable. For example, he slew the serpent Python in his lair. Python's lair was Delphi, the place which was said to be the center of the Earth.

Apollo, after the defeating of Python, took his lair for his own. Later, it would go on to be the Temple of Delphi. Delphi was where Apollo took on the role of the god of prophecy.

Delphi had an underground cavern, said to be a connection to the Underworld where vapors seeped out. These vapors were said to induce its inhaler to have visions regarding the future. A priestess of Apollo would be chosen as the 'Pythia' who breathed in the gases and speak the future in place of the god Apollo.

Apollo's most famous myth is the myth of him and Daphne. This myth begins with Apollo insulting Eros, the immortal son of Aphrodite, and also the god of passion. Eros's main weapon for causing havoc with love was his bow and arrows. The gold arrows would fill the target with love for the first thing that they looked at while the lead ones would fill them with dislike and hatred.

Apollo would insinuate that Eros, with his young body form, was a lesser archer than him. Eros, taking offense at the insult bided his time until he spotted a pretty nymph, Daphne, who was simply in the wrong place at the wrong time.

Eros made Apollo fall in love with the nymph with one of his golden arrows while hitting Daphne with one of his lead ones. Apollo started chasing after Daphne after she refused his initial advances towards her and ran away.

When Apollo finally caught Daphne at the edge of a river, she cried out to Gaea, the earth goddess to help her escape from the god's clutches. Gaea took pity on her and transformed her into a laurel tree, protecting her from Apollo.

In the ancient Olympics, winners would be awarded the laurel wreath. The wreath would have its bases in this myth, celebrating victors on their winning of the impossible. The laurel wreath would be symbolizing Daphne and the unwinnable.

One of the other main myths that feature Apollo is the myth of his son Phaethon. Phaethon was the son of Apollo. Apollo swore on the Styx to give him one wish that was in his power to grant.

Phaethon wished to drive the sun chariot for a day. Apollo allowed him to do so. However, Phaethon, only being mortal, lost control of the horses of the sun chariot. The horses went on a rampage. Forests burned, and chaos rained down on Earth. Some versions of the myth say that the heat of the sun was too hot that it burned the Africans' skin black.

Zeus, who as king of the gods, had no other choice but to strike Phaethon down with a thunderbolt. Phaethon would die and his death would be greatly mourned by his father.

I believe that the myth about Phaethon represents the sun well. While it warms us, humans, in the sky, it is dangerous and not to be messed around with.

While the ancient Greeks believed that the sun was just the sun god driving his chariot across the sky, we present-day humans know better. In school we learn that the sun is a big fiery ball of gas, serving as the energy source for the earth.

So, now that we have covered the Greek version of the sun, it is time to discuss our present-day view of the sun.

If we first look at the structure and composition of the sun, we learn that on the surface of the sun is the photosphere, chromosphere, and the corona, from the lowest ascending.

The photosphere is the part of the sun that we look at, the bright surface of the star. It emits energy in the form of visible light.

Then, we move up to the chromosphere. The chromosphere is the middle layer of the sun's atmosphere. It stretches on for about 2,000 kilometers. The chromosphere is also visible during a solar eclipse, a phenomenon we will be looking at later on.

After the chromosphere, we finally reach the corona, the outer layer of the sun's atmosphere. The corona extends many thousands of kilometers above the photosphere. While the corona of the sun is usually hidden by the bright light of the Sun's surface, we can see it during a total solar eclipse.

While the corona is dim, it is also very hot. This is due to the fact that the corona is less dense, about 10 million times than the surface of the sun. The sun has a magnetic field, which we will also discuss later on, and it is the magnetic field that causes charged particles to form shapes, such as loops and plumes.

The corona is special for one other reason. The solar winds that astronauts talk about are actually an extension of the corona. Because the corona has such a high temperature, its particles move at tremendous speeds, with a great deal of energy. The speeds that they reach are so high that the particles are capable of escaping the sun's gravity, therefore making the solar wind.

Now we move on to other features of the sun. These include sunspots, granules, solar eclipses, and much more.

Sunspots are dark spots on the photosphere. They are much cooler than the rest of the photosphere's temperature. This is mainly because they form over regions of intense magnetic activity.

Sunspots are almost 2000 degrees cooler than the surrounding brightness. While the photosphere has a temperature of 5,800 degrees Kelvin, sunspots only have a temperature of about 3,800 degrees Kelvin. They are also very large and can be up to 50,000 kilometers.

One other fact we need to know about sunspots is that they are able to generate mass eruptions on the sun. This is because sunspots are caused by interactions with the Sun's magnetic field. When this energy is released, solar flares and coronal mass ejections happen.

Early astronomers learned many things from observing these sunspots. They found that the sunspots move from east to west. The early astronomers also observed that the sunspots move faster if they are closer to the equator of the sun and slower at its poles. However, while it may look that way from earth, in reality, the sunspots are stationary while the photosphere moves.

This means that the sun rotates from west to east. It also implied that the sun was a gas. Because if the sun were solid, all the sun's sunspots would move in a straight line, rather than the ones closer to the equator being faster.

In the 19th century, scientists also found out that the number of sunspots on average go up and down every eleven years. This is due to the Sun switching poles every eleven years. This means that the Sun's north and south poles switch places and it takes another eleven years for the poles to flip back again.

If the number of sunspots increases, it means that the magnetic activity inside the sun increases and so the sun's magnetic field gets larger.

There are a lot of phenomena that could occur when

the sun has a lot of activity. The solar wind gets larger, and because the particles inside the solar wind collide with the earth's atmosphere, many things happen.

First, the Northern and Southern Lights happen more often. This is because the aurora occurs due to the particles of the solar wind creating friction with the earth's atmosphere. The Northern Lights and Southern Lights increase so much to the point that in the highest sunspot year, 1859, they were visible from as far from the poles as Cuba and Hawaii.

The other phenomena that could happen are things such as impacting radio communications and messing with airplane navigation technology. If solar activity is high enough, it could even affect electricity grids on Earth.

Next up are granules. Granules form on the surface of the sun and they resemble tiny grains of rice, hence their name.

Granules form due to convection currents of plasma within the Sun's convective zone. The hot part of the granules is located in their center, while cooler regions are on the outside.

Since luminosity is in proportion to the fourth power of temperature, even a small difference in temperature may cause different colors. The golden parts are the hotter parts, while the black in between them are the cooler ones.

Then the next topic is the solar eclipse. A solar eclipse happens when the Moon gets between the sun and the Earth and prevents sunlight from reaching the Earth. It means that the moon moves over and covers the sun, making it dark.

A solar eclipse only happens when the disk of the Moon appears to cover the disk of the sun.

However, because the moon's shadow is not very big, only a few places in the world can see a total solar eclipse. But there is also a partial solar eclipse. In a partial solar eclipse, the whole of the sun won't get covered by the moon, only part of it.

The difference between a total solar eclipse and a partial eclipse is that while in a total solar eclipse, the Earth, moon, and sun are perfectly aligned, in a partial solar eclipse, they are not perfectly aligned and so the moon does not cover the entirety of the sun in the sky.

Chapter Twelve :

Artemis

Whoso is chaste of spirit utterly
May gather leaves and fruits and flowers.
The unchaste never.

02

Artemis, like her twin brother Apollo, was the offspring of Zeus and the Titaness Leto. She was the goddess of the moon, the hunt, childbirth, and chastity. Like her brother, Selene was the original goddess of the moon before her, but Artemis replaced her.

Similar to her brother, Artemis was also very vengeful against those who had offended her mother Leto. One example of this is when she and Apollo slew the queen of Thebes's royal family.

Niobe, the then queen of Thebes, had given birth to fourteen children. Seven of them were sons while the other seven were daughters. Niobe was arrogant and boasted that while Leto had only managed to birth two children, she had fourteen of them.

Understandably, the twins were outraged at Niobe and proceeded to slay them. Although the seven sons and Niobe's husband, the king of Thebes, attempted to defend themselves, they were slain by Apollo's bow. The seven daughters were slain by Artemis.

Niobe was so overcome with grief that she went into seclusion in a mountain and stayed still in one spot with tears streaming down her face. Legend has it that she was in the position for so long that she was petrified and turned to stone. However, she still has water leaking from her unseeing eyes.

Artemis, along with her role as the goddess of the moon, was also the goddess of the hunt. She had a group of handmaidens who would hunt alongside her and assist her when she was recuperating from hunting.

With her being a maiden goddess, only pure maidens could provide her with companionship. However, many gods and humans would attempt to rape her handmaidens or even Artemis herself. Their results are the myths below.

One myth featured a hunter named Actaeon. He

accidentally saw Artemis and her handmaidens bathing in a spring. However, he did not leave or beg for mercy right away but tried to force himself on her.

For his punishment for seeing her and her handmaidens' nude forms, Artemis turned him into a deer, and Actaeon was subsequently torn apart by his own hunting hounds.

Another myth is the myth of the Aloadae. The Aloadae were twin sons of Poseidon who grew enormously and could not be killed except by the other's hand. Their growth never stopped and they attempted to stack mountains on top of each other to reach Mount Olympus. They asserted that they would kidnap Artemis and Hera for their wives.

Artemis turned herself into a fine doe and ran between the two giants. The twins, each seeking to claim the kill of the doe as their own, threw their spears at it. But to their misfortune, Artemis was simply too fast and the spears ended up impaling each other instead.

However, the goddess was not all that harsh to men in general. She even took a hunter named Orion as her hunting companion and allowed him to join her in her hunting.

Sadly, Apollo was a little too protective of his twin sister and ended up tricking Artemis to kill him with her hand. Apollo was jealous of his sister's love towards Orion and worried that she might forget herself. Apollo tricked Artemis into believing that Orion in the distance was a target and goaded her by stating that he could shoot more accurately than she could.

As Apollo had planned, his sister did not miss the shot and Orion died. Artemis was distraught over the death of Orion and made a constellation of him in the sky to remember and respect him.

However, gods also targeted Artemis's handmaidens. An example is Callisto. Being an attendant of Artemis, she, like the goddess, had taken a vow of chastity. When Artemis was away, Zeus disguised himself as Artemis and raped Callisto. Despite Callisto trying to conceal the fact that she was pregnant, she was discovered after refusing to bathe with the others.

Although Artemis pitied Callisto, she had defied the goddess and defiled her fellow handmaidens with her presence. The goddess of the hunt punished Callisto by turning her into a bear.

Later, Callisto gave birth to Arcas. He became the king of Arcadia and was a fearsome hunter. When he was out hunting, Callisto, in the joy of seeing her son, who she had not seen for many years, tried to go to him.

Arcas, not knowing that Callisto was his mother, took out an arrow to kill the bear. However, Zeus interfered before Arcas could do so and turned the two of them into constellations, specifically the Ursa Major and Ursa Minor, the little and big bears.

While the Greeks believed the goddess Artemis was responsible for the moon, we present-day humans have a widely different perspective of the moon.

The moon is the brightest and largest object in our night sky. It is Earth's only natural satellite. The moon, while being smaller than our Earth, still has a lot of influence on our planet, especially oceanic organisms. An example of the moon's influence include the tides.

There is also some information that we need to know about the moon. This includes the moon's phases and the lunar eclipse. First, the tides are the effect of the moon's gravitational field on the Earth. Because the moon is smaller in mass than the Earth, its gravity does not have a big effect on the land, but the seas are a different story.

Although the moon cannot affect the land, it can affect the water, and its gravity is what causes the high and low tides. As the moon's gravity pulls at the Earth, it pulls the water closest to it and the water furthest from it into two distinct bulges, one close to it and one on the opposite side of the moon.

The bulge of water closest to the moon is caused directly by the moon's gravity. The other bulge forming where the Earth and the moon are furthest apart would seem a bit strange.

The other bulge is due to inertia, the principle that unless an outside force exerts energy on an object, the object will maintain its current state of motion, be it moving or stationary.

The gravitational attraction of the moon is smallest on the far side of the Earth, and because of this, inertia exceeds the lunar gravitational force, and the water tries to keep its original state of motion by moving in a straight line, hence forming a bulge.

So, to summarize, the moon affects the Earth's oceans with its gravity. The moon forms two bulges of water. The first bulge is made by pulling the water closest to it. The second bulge forms on the opposite side of the Earth from where the first bulge formed. This is because of the force inertia. By forming the two bulges, the moon forms the high tides and the low tides.

The lunar eclipse is when the Earth's shadow obscures the moon. Lunar eclipses happen at the full moon phase. During a lunar eclipse, the Earth gets in between the sun and the moon and passes its shadow over it.

There are three types of lunar eclipses. The first is the total lunar eclipse. The total lunar eclipse occurs when the moon fully moves into the inner part of Earth's shadow, called the umbra. However, unlike the solar eclipse from earlier, the moon does not disappear with its journey into the umbra. Rather, it turns red.

The reason behind this phenomena is light refraction. When the moon passes into the umbra, light from the sun isn't able to reach the moon. However, light from the Earth's atmosphere can still reach it.

Blue light is scattered by the atmosphere comparatively more than red light. Although all light of different wavelengths travel at the same speed, blue light has a higher frequency and is shorter. So, it is refracted and scattered comparatively more than red light, which has a lower frequency and is longer.

This means that only red light is able to reach the moon and reflect back to Earth successfully. This is the reason why when the total lunar eclipse happens, the moon turns red.

The second type of eclipse is the partial lunar eclipse. Unlike the total lunar eclipse, the partial lunar eclipse only passes through part of the umbra. The shadow of the Earth grows, but recedes without entirely swallowing the moon.

The third and last type of eclipse is the penumbral eclipse. The penumbral eclipse happens when the moon only travels through the penumbra, the outer part of Earth's shadow. Because the penumbra is faint, the moon only dims slightly.

Chapter Thirteen :

Ares

Ares, exceeding in strength,
chariot-rider, golden-helmed, doughty in heart,
shield-bearer, Saviour of cities, harnessed in bronze,
strong of arm, unwearying, mighty with the spear,
O defence of Olympus, father of warlike Victory,
ally of Themis, stern governor of the rebellious,
leader of righteous men, sceptred King of manliness,
who whirl your fiery sphere among the planets
in their sevenfold courses through the aether
wherein your blazing steeds ever bear you
above the third firmament of heaven

Ares was the son of Zeus and Hera and the god of war. Unlike his sister Athena who represented the strategic and logical aspects of war, he was the personification of the more primitive state of war, representing the more emotional parts of war.

Ares was, as the Iliad so eloquently put, the most hated god on Olympus. Even his parents and fellow gods were not very fond of him. For example, when Ares got hurt in the Trojan War, Zeus swore that he would throw him off Olympus if he had not been his son. However, he did have his followers in the form of Amazons, Spartans, and other people.

The Amazons were a group of women warriors in

a land far away from Greece. They were a cult of women who matched men in physical prowess and battle. Their society was very enclosed and exclusive only to their female children; the males were cast out and left to die. The Amazonian warriors worshiped Ares as their patron god.

Another society that worshiped the war god was Sparta. Known for its harsh military training and strict disciplinary rules, it was one of the two most powerful city-states, the other being Athens. The Spartans were said to have tied a statue of the god in the middle of their city so that victory in war would never leave them and always be assured.

Ares rode into battle with his two charioteers, his sons Phobos and Deimos, both being gods of fear. Phobos inspired panic and flight in his enemies while Deimos was the representation of fear and terror.

The myths that include Ares all take the same perspective of him, showing him to be a violent, bloodthirsty, arrogant, raging fool. One such example is when he was kidnapped by the Aloadae.

As seen in the previous chapter, the Aloadae were

twin giants, named Otis and Ephialtes planning to overthrow Mount Olympus and take the goddesses Artemis and Hera for their wives.

While the giants are eventually defeated by Artemis, by means of trickery, Ares had first tried a brunt assault against them. However, he was overpowered and locked inside a bronze jar as a prisoner for thirteen months.

It was not until the Aloadae's mother, Eriboea, came to the god of travelers, Hermes, and told him where Ares was being kept that Ares was rescued. After Ares was rescued, the twin giants killed each other thanks to the trickery of Artemis, as seen in Chapter Thirteen.

Another myth about Ares shows him in a more forgiving light with the King of Thebes. Cadmus was the first king and founder of the city of Thebes. Cadmus was actually the brother of Europa, who was abducted by Zeus in the shape of a cow, as seen in the earlier chapter.

Cadmus's journey started at the Oracle of Delphi. Upon consulting the oracle, Cadmus was instructed to give up the search for his sister and to follow a special cow and build a city where it stopped to rest.

When the men looked for a well to quench their thirst, they found a water dragon sacred to Ares protecting it. Athena, the goddess of wisdom and war, advised Cadmus's group to kill the dragon.

The goddess then instructed the men to sow the dragon's teeth into the ground. From the sowed teeth sprang a legion of strong warriors, called the Spartoi. The Spartoi would help in protecting the town from outside invaders.

Ares, however, was displeased with Cadmus for defeating his sacred dragon and had Cadmus serve 8 years of penance for his crime against him. After his punishment, Ares forgave him and chose to give Cadmus his daughter Harmonia for his wife as a gift.

Cadmus's suffering would not end there, however. On his wedding day to Harmonia, they received a cursed necklace which brought the wearer extremely bad luck. So bad that it in fact was able to turn Cadmus into a serpent after his whim that the life of a serpent would be better than his if the gods took such an interest in it.

Harmonia begged the gods to be able to share

Cadmus's fate. The gods taking pity on her, granted her wish. The two serpents would wander off into the wilderness. In the end, Zeus, taking pity on them, turned them back human and transported them to Elysium, where pious souls go after death.

Another myth about Ares was when he got tried with a crime. The myth begins with Poseidon's son Halirrhothius raping Alcippe, Ares's daughter. When Ares found out about Halirrhothius's crime, he killed him in a fit of rage.

Halirrhothius's father, Poseidon, was angry at his son's death and had Ares put on trial at a hill near the famous Acropolis. Today, we call this hill the Areopagus, roughly translated as 'Ares's Hill'.

Ares was acquitted for his act of felony, and later, the ancient Greeks used the Areopagus as a place to trial their own criminals, mainly those accused of homicide.

Ares was the god of war. In the ancient world, catapults with flaming incendiaries were used as weapons. One of the most famous flaming weapons used in history is Greek fire.

To know about it, a history lesson is in order. When the Roman Empire spilt into their eastern and western halves, the western version was conquered by Germanic tribes. However, the eastern version of Rome survived to form the Byzantine Empire.

Greek fire, often called sea fire in ancient texts, was a weapon that set fire to enemy ships in naval battles. It consisted of a combustible compound thrown by a flame emitting weapon. Greek fire was developed and used as a weapon in war by the Byzantine Empire, in wars to protect itself against invasions from the Arabs.

Historians believe that the mixture of Greek fire was made from quicklime and naphtha. Quicklime is another name for the chemical compound calcium oxide or CaO. Naphtha was a substance made from the fractional distillation of oil. The two chemical properties of quicklime and naphtha made for an effective incendiary.

Quicklime, or calcium oxide produces an exothermic reaction if it comes into contact with water. In the case that it does, the calcium oxide will give off an extreme heat, turning water to steam. In naval battles surrounded with water, calcium oxide would be the perfect material to set enemy ships aflame.

Exothermic reactions happen because of the difference between the reactant and the product of a chemical reaction. In a chemical reaction, bonds between different atoms and molecules are either formed or broken.

If bonds are formed in a chemical reaction, heat is released, and if bonds are broken, heat is absorbed. Particles are more stable together than when they are separated. So, when bonds are formed between particles, it takes less energy when bonds are broken.

The reason that particles are more stable together than when they are apart is because of the octet rule. The most stable particles are inert gases, the chemicals of group 18 on the periodic table.

Particles want to be stable, so they share bonds with other particles in order to be like the inert gas that is closest to them. They do this by filling the first electron orbit with two electrons and the rest with eight electrons.

For example, if we take table salt (NaCl), sodium (Na) has a single electron on its outermost electron orbit. Chlorine (Cl) has seven electrons. The sodium particle gives its electron to the chlorine particle. This way, the sodium can have the same electron configuration as the inert gas neon (Ne) and the chlorine can have a configuration like argon (Ar).

Some other exothermic reactions like the one with quicklime and water include the famous reaction of hydrogen and nitrogen to form ammonia.

Naphtha is made from the fractional distillation of crude oil. Fractional distillation works by using the difference in the boiling point of materials to separate them. When a substance has a lower boiling point than the rest of the materials in a mixture, it separates from the whole as a gas. This gas is then collected and cooled to form a pure liquid substance.

Naphtha is separated at a temperature of 120 degrees. It has two different uses. The first one is fuel. Naphtha has a low boiling point at about 35 degrees Celsius, which makes it easy to ignite. Although it forms into a gas at about 120 degrees, as seen earlier above, it starts its vaporization at 35 degrees.

The second use of naphtha is in the production of plastics. Naphtha is a crucial ingredient in the process of making plastics.

With the sparking combination of quicklime and naphtha, it is no wonder that the Byzantine Empire survived a thousand years longer than its western counterpart.

Chapter Fourteen :

Hephaestus

Thrown by angry Jove

Sheer o'er the crystal battlements; from morn

To noon he fell, from noon to dewy eve,

A summer's day, and with the setting sun

Dropt from the zenith like a falling star,

On Lemnos, the Aegean isle.

02

Hephaestus was the god of metalworking, forges, and fire. Hephaestus's life would start on a heavy note. Unlike the other gods, he had no father. But he had a mother, Hera, the queen of the gods. Because Hera was tired of her unfaithful husband, she conceived a child on her own without any male help. However, when Hephaestus was born, he was so ugly that Hera, disappointed, threw him out the window.

Legend has it that it takes a total of nine days to fall from Olympus to Earth. Hephaestus fell all nine days down. He eventually landed in the ocean where the goddess Thetis cared for his injuries and raised him.

When Hephaestus grew up, he journeyed back to

Olympus to confront his mother about what she did and to get revenge on her. Hephaestus's revenge came in the form of a present for Hera a throne that he had made himself.

The throne was so beautiful and masterfully crafted that Hera accepted the gift without any reluctance. But Hephaestus had enchanted the throne to tie Hera up so she could never escape its clutches without his permission.

The other gods tried various plots and schemes to get Hephaestus to release Hera. However, nothing seemed to work. That was when Dionysus, the god of wine came into play.

Dionysus offered Hephaestus his wine and, in no time, the god of forges was intoxicated. In his drunken state, Hephaestus was easily persuaded to let go of his anger and release Hera.

Hera and Hephaestus would eventually work things out and learn to forgive each other and move on. Hephaestus would be caught in trouble, not for acting against his mother but for helping her.

When Hera was caught by Zeus to be one of the instigators of a rebellion against him, as punishment,

Zeus tied Hera over the entrance to Chaos, where, on the occasion that she fell, would fade back into nothingness.

Hephaestus felt sorry for his mother and got her down safely. Needless to say, Zeus was angered with his actions and Hephaestus was thrown off Olympus. Again. This time he ended up on the island of Lemnos, where the villagers helped him until he healed from his wounds.

All his suffering would not end there, however. When Aphrodite, the goddess of love and beauty (who is in the next chapter) arrived at Olympus, he would be selected by Hera to be her husband.

However, Hephaestus was not good-looking. His mother threw him off Olympus for his ugliness and two falls to Earth could only have worsened his looks. So, Aphrodite was not attracted to him at all. Instead, she started an affair with Ares, the god of war. Aphrodite and Ares would even go on to have children together, including Harmonia, and Eros.

Hephaestus would remain clueless about his wife's affair until Helios, the god of the sun, informed him about seeing the two Olympians together on his daily trip across the sky.

To see the affair for himself, Hephaestus set up a trap for the two gods. After putting up the trap, he made it look like he was going off on a long journey. Not much time had gone by when the trap was sprung. A golden net had trapped Ares and Aphrodite to the bed, preventing them from escaping.

Hephaestus called the other gods to see the spectacle for themselves. Then, after he had thoroughly embarrassed them, he let the two gods free, due to the objections of the other gods.

Hephaestus was a master craftsman and many of the other gods and legendary heroes came to him for weapons, equipment, and more. Arguably his best work amongst all the pieces that he made was Talos.

Zeus, the king of the gods, requested Talos to guard the island where the princess Europa who he kidnapped was living. Talos was an automaton, similar to the machines and AI that we have today.

It was powered with ichor, the blood of the gods, running through its veins. It could act autonomously and its single mission was to destroy invaders approaching the island.

However, Talos would fail, due to a group of invaders, more specifically Jason and the Argonauts. Their ship was hidden in a sea cave, away from Talos's sight.

However, on his daily rounds, he saw it. Medea, the witch who had fallen in love with Jason, realized Talos's source of energy, and in order to trick him, she cooked a devious plan. When Talos tried to destroy their ship, Medea made a deal with the automaton.

She offered to make Talos immortal. Because Talos did not know the mechanical nature of his own existence, Talos agreed, not knowing the darker intentions of the witch.

Medea sliced open Talos's metal veins for her immortalizing "ritual" and slowly, Talos's strength and life force drained out of the massive automaton.

It is interesting to see that the ancient Greeks had ideas of self-thinking autonomous machines and expressed their curiosity with fantastic myths. With the myth of Talos, the lines between robots and humans are blurred.

While Talos is somewhat misrepresented as the monster of the myth, I believe that he was the victim of his own story.

Talos lost his unknown immortal ichor-given life, due to him thinking that he was mortal, like the rest of the humans he saw. Because we are actually mortal, we can imagine the temptation of immortality. Who amongst us wouldn't want to exchange our mortal lines with an immortal one?

Because Talos also thought along this path, we as the readers as well as being truly mortal can sympathize with the automaton's desire. This blurs the boundaries between humans and robots.

That is what is happening with the Fourth Industrial Revolution right now. The lines between robot and human are becoming non-distinguishable, and in the future, will be nonexistent. This change is happening under the banner of the field of artificial intelligence, or AI paired with the field of machine learning.

There are three main ways of training AI. Each of them is called supervised learning, unsupervised learning, and reinforcement learning, respectively.

Let's look at supervised learning, to begin with. Supervised learning is when an algorithm is trained with pre-expert labeled datasets, with the objective of learning to classify data or predict accurate outcomes.

As the input data rises, so does the algorithm's accuracy. The algorithm adjusts its determining process with each labeled data that comes its way. It figures out what formula of determination yields the highest accuracy.

An example of this supervised learning is with an algorithm learning to distinguish between cats and dogs. Although this may seem like an easy problem for us humans, we owe our ability to differentiate between the two species to hundreds of thousands of years of evolution and survival.

So, for a trained algorithm to match those hundreds of thousands of years of training with just calculations such as what the angle of a cat's ear is, is sort of like the golden tier for algorithms.

With the example of separating the cats from the dogs, the algorithm examines a data set, previously determined by a human expert. It then recognizes specific characteristics between the same classification in a dataset and based on those similarities, makes its decision on how new data should be labeled.

In supervised learning, the trainer of the algorithm helps the algorithm out and corrects it when it gives the wrong answer.

Second up is unsupervised learning. The difference between supervised learning and unsupervised learning is that the latter uses unlabeled data.

Unsupervised learning is a self-learning technique where the algorithm itself has to find similarities between numerous data of a given set and separate the dataset according to those or other similarities.

This type of AI learning is when we are unsure about a given set of data ourselves. There are no prelabeled datasets, only the algorithm finding patterns in a sea of knowledge.

One fun application of unsupervised learning is when we use it to look for alien signals in the radio waves that pass by Earth. In the VLA- which stands for Very Large

Array, the radio dishes combine to form, in operation, one giant radio dish. The signals are then sent to a research site.

Researchers then look for repeating patterns within the signals, and they use algorithms with unsupervised learning to do so. Because it is beyond the abilities of humans to look for every single thing that may or may not be repeating, the unsupervised learning algorithms do the work.

Although the signals that we have translated now are equal to a glass of water compared to the entire amount of water in the oceans, by harnessing the power of AI and machine learning, I believe we can find a signal from extraterrestrial life out there somewhere in our vast universe.

Then finally we move on to reinforcement learning. In reinforcement learning, the specialist models the algorithm in such a way that it is able to interact with the environment and know what was productive while another was not.

The best metaphor for reinforcement learning is a gambling bar. However, unlike the gambling bars in real life, each game reveals a consistent reward. There are many games in the bar and you don't have time to try them all, and you must reach a certain number of rewards.

What choice you have is divided into two parts. They are called exploration and exploitation. Exploration is when you spend your time exploring different types of games, along with the number of rewards that they give you.

The second option is that you exploit the game that, through your exploration, have found that yields the greatest number of rewards out of all the games that you have tried so far.

One example of reinforcement learning in real life is the Netflix algorithm. It works with the purpose of enticing people to continue to pay for its service, which is showing movies. However, it must first know what type of seduction works best on the audience.

Then it spends its time exploring and exploiting different ways to make people open their wallets another time. This example is related to the movie thumbnails. Netflix's algorithm is tricky. It is different for every user.

According to the user's screen and skipping scenes

of a movie, the algorithm gathers its data. Then, it shows a thumbnail on a page of thumbnails, which, based on their user's stopping and skipping the parts of a movie, which should entice the watcher to click on the movie.

According to the number of times the user clicked on the newly designed thumbnails, the algorithm will decide if its new designs were successful. If it was, it would exploit that new design until enough information is gathered for the next design of thumbnails.

Let us go back to the original topic of the myth of Talos and discuss another story that has the implicit moral lesson as the myth of Talos. This story is <The Bicentennial Man> by Isaac Asimov. However, the difference between the myth and the story is that in <The Bicentennial Man> the robot gives up his immortal life as, well, a robot, in favor of one as a human.

So, I guess opinions on immortality may vary from person to person, and with all the collected experiences in our minds, after a while, we may get bored of living.

However, in our experience, our allotted century is not nearly enough. So, what we can do is refer back to the chapter of Hades and look at the mind uploading technology.

People could exchange their organic brain in favor of a digital one. In their digital selves, they could organize their affairs, and let go when they feel it is the right time for them to do so.

Along with Talos, Hephaestus also made Pandora, along with the help of a few other gods. Because Zeus wanted to punish the family of Prometheus for Prometheus gifting humans fire, he ordered Hephaestus to make Pandora.

Zeus wanted to punish humans as they are Prometheus's creations. Zeus made them suffer illness, death, madness, and more. Before the gods' tricking of Pandora, life had been good for humans, them knowing no suffering.

Pandora was to be the gods' weapon against Prometheus's brother, Epimetheus. Although Prometheus had warned his brother to beware of gods bearing gifts, if Pandora became his wife, and the gods sent her a gift, he would have no choice in whether to accept the gift or not.

Because Hephaestus had crafted Pandora so finely

and Aphrodite herself had bestowed grace and beauty upon her, along with other such gifts from the other gods, Epimetheus stood no chance.

Epimetheus and Pandora were engaged soon enough and that was when Zeus struck. He, along with the other gods, had blessed or cursed, depending on how one looks at it, Pandora with curiosity.

At their wedding, Zeus sent Pandora a jar filled with all the evils that humankind had yet to know. It was filled with pain, suffering, sickness, and the like. When the gods gave the jar to Pandora, they told her they were entrusting it to her and warned Pandora to never open it.

However, because of Pandora's curiosity instilled in her by Zeus, it was only a matter of time before Pandora opened the jar. When she opened the jar, while she let all the evils out, she slammed the lid down before one spirit could make its escape.

That spirit was Hope. Since then, no matter what hardships humans have faced, we have always been able to cling to the last spirit, hope.

In a twisted sense of irony, maybe Pandora's jar

was not actually a curse on all of humanity but rather a sort of wake-up call for us. Because we did not know what suffering was, we could not appreciate how blessed we once were.

But now, because we have our suffering as a measure of how bad things can get at one end of the spectrum, at the end of the day, we also know how good things are at the other end. It is rather like the adage, the one that states that there can be no light without darkness.

Chapter Fifteen :

Aphrodite

The breath of the west wind bore her
Over the sounding sea,
Up from the delicate foam,
To wave-ringed Cyprus, her isle.
And the Hours golden-wreathed
Welcomed her joyously.
They clad her in raiment immortal,
And brought her to the gods.
Wonder seized them all as they saw
Violet-crowned Cytherea.

02

Aphrodite was the goddess of love and beauty. Her sacred animal was the dove and her symbols were the apple, goose, and lettuce. She had no parents and her origins were a little stranger than those of the other gods.

When Kronos overthrew Ouranos by slicing him into pieces with the scythe that his mother Gaea granted him, one of the pieces of Ouranos's body fell into the ocean. Aphrodite rose from the seafoam when the body part collided with the sea.

Much like the nature of her birth, Aphrodite was savage. She was used to getting what she wanted, just because of her looks. She even had a girdle of love that her husband Hephaestus forged her, which would make her look more attractive to everyone.

Some myths that include the darker side of love are when she collaborated with Poseidon to punish Minos and Pasiphae, the king and queen of Crete.

Minos had prayed to Poseidon to give him a sign that the gods favored him over his two brothers in order to win the crown of Crete, and, seeing that Minos was the best choice for king, Poseidon had sent a pure white bull to Minos, on the condition that it be sacrificed to him after it had served its purpose.

However, Minos had not sent the original white bull, but another, thinking that he could trick the sea god and keep the pure white bull for himself. And so, Poseidon was angered with Minos.

Pasiphae had angered Aphrodite by means of claiming that she was much more beautiful than the goddess.

Aphrodite had made Pasiphae fall in love with her husband's prize bull gifted from Poseidon. Eventually, Pasiphae gave birth to a half-human, half-bull monster called the Minotaur.

Because of its bloodthirstiness, the master inventor Daedalus would create an unescapable maze, the Labyrinth, for the Minotaur to be locked in.

Minos would also look for ways to quench the Minotaur's hunger and he settled on making a then weaker Athens send seven young men and women for them to be a sacrifice for the beast in the maze.

Then one year, the prince of Athens, Theseus, demanded to be sent as a sacrifice in order to end the Minotaur once and for all. When he arrived at Crete, princess Ariadne fell in love with him at first sight, as did Theseus.

On a promise to take her home and marry him, Ariadne agreed to help Theseus kill the Minotaur. Her help came in the form of a sword because all of the fourteen youths were to be unarmed, along with a ball of string, to follow to come back to the entrance of the maze after he had done the deed.

While Theseus was successful in killing the Minotaur, he went back on his promise to take Ariadne back to Athens and marry him. So, he ditched her on an island while she was sleeping on his way back home, Naxos.

However, Ariadne wasn't going to be left on the island forever. The god Dionysus would see her alone on Naxos, fall head over heels for her in love.

However, much like the dual nature of love, the arena in which Aphrodite reigned supreme, she also had her forgiving and soft side as well. One of the myths that show this part of the goddess is when she gave life to a stone statue.

The myth begins with a statue carver in the city of Cyprus named Pygmalion. Pygmalion loved his work and decided to spend the rest of his life devoted to carving, and swore off love forever.

However, one day, Pygmalion began carving an ivory girl, one more beautiful and purer than the shallow women in his town. He made a statue of a woman that was so perfect, he fell in love with it.

Pygmalion named the statue Galatea and lavished upon it gifts that he bought for her. He dressed her in the finest silk, decorated her arms with expensive jewelry, and expressed his love and devotion to her every day.

However, she did not talk back. Pygmalion fell into

misery and hopelessness. His only hope was the annual festival of Aphrodite, the patron goddess of his city Cyprus.

When the day of the festival finally came, Pygmalion prepared the sweetest apples, the symbol of the goddess of love, and when it was his turn to pray for love at the altar, he was too embarrassed to admit his love for his ivory statue in front of all the townspeople and he replaced his wish for one to find a woman as beautiful and pure as his statue.

Unbeknownst to Pygmalion, Aphrodite had heard his wish from afar on Olympus and knew what he really wished for. Aphrodite thus granted Pygmalion's desire.

After the festival, when Pygmalion went home, he went and looked at Galatea for a long time, hoping for a miracle. However, when nothing of the sort happened, he was devastated.

Pygmalion went to destroy the statue, thinking that maybe he wasn't destined for love at all. Before he dealt the statue a blow, he gave it one last kiss. During the kiss, Galatea's cold, hard stone face turned soft and warm. Aphrodite had granted his wish and turned Galatea into a human.

The most interesting part of myths pertaining to Aphrodite was the one with the Minotaur. Could something like that actually happen?

Although all types of animals are just different branches of the Tree of Life, no two different species can produce viable, fertile offspring.

This is mainly due to the fact that once different branches separate from each other far enough, they are not of the same race and therefore are unable to produce offspring.

However, there are exceptions to this rule. For example, a female horse and a male donkey are able to produce offspring: the mule. But, while the horse and the donkey may be close enough in race to bear offspring, they are different enough that the offspring produced is infertile. This is the reason that mules are not able to produce offspring.

This fact is also related to Pasiphae, the queen of Crete, birthing the Minotaur, a monster sired by a white bull. This should not be able to happen in real life.

Because, while the horse and donkey are close enough in race in order to produce the mule, we humans and cows are definitely not close enough to birth even an infertile child.

So, unless in the time of the ancient Greeks, we were closer to cows than we are now, the myth of the Minotaur would have to be put down to mere fantasy.

Chapter Sixteen :

Hermes

The babe was born at the break of day,
And ere the night fell he had stolen away
Apollo's herds.

02

Hermes was the god of travelers, thieves, and much more. His symbols were the lyre, along with his caduceus. The god of thieves started his adventure when he was only a babe.

Hermes is often represented with winged sandals because he was the messenger of the gods and needed to move fast in order to deliver messages between them.

Hermes was born with Zeus as his father and the mountain nymph Maia as his mother. His first myth and adventure starts when he was only a baby. He and his mother Maia lived in a cave in order to hide from Zeus's wife Hera in fear that she would rain punishments on them.

After his mother Maia had gone to sleep, Hermes

snuck out of the cave and went to find adventure. What Hermes first found was a tortoise wandering just outside the cave. He killed the tortoise and stretched strings of sheep gut over its empty shell. Hermes had created the first lyre.

After teaching himself to quickly play the instrument that he had invented, Hermes resumed his adventure. Hermes had gone a long while when he spotted the pasture of the gods.

Because Apollo, the then herdsman of the gods, was not there to watch the cows, Hermes, even as a young god being just as cunning and tricky as his older self, disguised his identity by wearing large shoes designed from leaves and led fifty of the herd away.

Olympians and ate some of the meat. The rest, however, he led into a cave and hid them. Then, he went back to the cave he lived in with his mother in.

On the way back home, Hermes was spotted by a farmer by the name of Battus. Hermes bought the farmer's silence about him and returned to the arms of his mother.

When morning came and Apollo came to check

on the cattle, he immediately noticed that fifty of their number were missing. He also noticed the giant footsteps Hermes had left behind as a ruse to throw him off his scent. While he searched all over the world, he could not find the missing cows and their supposed giant kidnapper.

Then, when Apollo was searching the nearby countryside, he ran into Battus, the farmer. Although Hermes had already bought his silence, the man's loose tongue talked about everything he had seen that early morning, including the babe he had seen walking on the road alone.

Thanks to Battus, Apollo traveled to the cave Hermes and his mother Maia had claimed as their home.

Hermes was initially surprised at the god's appearance; he had not expected the god to identify him as the thief that quickly. However, the young god recovered quickly, acting innocent, with the help of his mother, who was kept in the dark and ignorant about his actions.

But eventually, Hermes slipped up and revealed that the number of stolen cows was fifty, while Apollo had never told him the exact number. Apollo, having caught onto

Hermes's ruse quickly, took Hermes to Olympus to be tried for his crime, planning to force him to reveal where he hid the cows that he stole after.

When the two gods arrived at Olympus, Zeus and the other Olympians decided that while Hermes had committed a crime, he had sacrificed some of the cows to Olympus and instead of giving Hermes the harsh punishment of throwing him into Tartarus, decided to let Apollo give him a punishment of his choice, albeit one kinder than that of throwing him into the pit of eternal damnation, Tartarus.

After the conclusion of Hermes's trial, Apollo decided to first-order him to take him where he hid the stolen cows.

On the way to the hidden cows, Hermes decided that the best way to get out of the situation was to barter with Apollo. The only thing that Hermes had to offer was his lyre, the instrument that he had invented the night before, hollowing out a tortoise's shell and attaching cow guts for strings.

Apollo, being the god of music, saw the potential

that the newly invented lyre had and was desperate to get possession of the lyre. Hermes, seeing a chance for further room to trade, asked if Apollo could offer him anything along with his freedom.

The god of music offered the inventor of the lyre a caduceus, some dice to predict the future, and the sharpest sword in the world, made of adamantine. Hermes gladly accepted the proposed offer.

Now, the lyre is one of Apollo's main symbols, where the caduceus is one of Hermes's. So, both gods came out of the offer getting what they wanted, although Apollo's side of the deal was a bit ripped off.

After the fiasco of the stolen cows, Hermes was made an Olympian, the god of thieves, travelers, and much more. His symbol was that of the caduceus, the same thing he had received from Apollo in exchange for his lyre.

Another myth pertaining to Hermes was the one of Io. Io was another one of Zeus's lovers. However, when they were meeting in secret from Hera, Zeus's wife, she found out where the two were meeting and rushed to the spot in order to confront them.

Zeus, thinking hurriedly, turned Io into a heifer, a female cow, in order to disguise her from Hera, thinking that it would hide Io from the force of Hera's wrath and revenge.

Zeus completed Io's disguise before Hera arrived at the scene. Then, he presented the cow to Hera, saying that it was a gift for her.

However, because of his harried thinking, Zeus failed to consider that cow was Hera's sacred animal. Upon seeing the heifer, Hera knew that there was something wrong with the cow and, acting on her suspicions, ordered her servant Argos to take the cow to her temple and guard her.

However, Argos was not a normal servant. He was a giant with a hundred eyes staring in every direction on his body. This ensured that Io was guarded and never got the chance to escape, be it on her own power or with the help of others. Also, the giant never slept.

Zeus called on Hermes to save Io from Hera's grasp, being that he was the god of thieves and had experience with stealing cows.

Hermes dressed as an ordinary traditional bard and

went up to Argos. The tricks up his sleeves included water mixed with a sleeping potion and his voice.

On a hot summer day, Hermes went up to Argos and gave him the water with the sleeping potion and as a bard, offered to play his lyre for him.

Although indulged with the sleeping potion, it was not easy to get him to sleep. Hermes droned on and on about the various epics of the gods and it was only a long time later that Argos finally fell asleep.

When the giant had found his way into the realm of Hypnos, the god of sleep, Hermes drew his sword and killed him. Then, he retrieved Io and freed her. In the end, Io escaped from Hera's grasp.

However, Hera still got her revenge on Io. She cursed Io with a gadfly to chase and bite her in her bovine form forever. Io then roamed the countryside, devoid of all hope, when she met a prophet who told her that she would wander for many years but would eventually be changed back into a human.

Io's travels eventually took her to Egypt, whereupon reaching the Nile river, Zeus changed her form back into a woman.

Hermes was the messenger of the gods, and he had winged sandals to move around quickly by way of flight.

However, under the assumption that gods need oxygen to metabolize and create energy like us humans, Hermes could only go upwards so far into the atmosphere before the low oxygen levels kicked in and made it impossible to travel.

First, to answer the question of why traveling higher is faster, because there is less air as one goes upwards in the atmosphere, there is less air resistance, which makes it a faster and more efficient way to travel.

We must first look at the parts of the atmosphere to determine how high a human (or god in this case) could go without any sort of oxygen-supplying device.

In order to do so, knowledge of the atmosphere is first needed. The atmosphere is made up of approximately four distinct layers.

The first layer is the troposphere. The troposphere stretches from the ground to about 11 kilometers up in the sky. In the troposphere, convection occurs actively. Also, clouds and other meteorological phenomena happen. This is because water vapor exists in the troposphere. If there weren't any water vapor, clouds would not form. Also, most of the air is located within the troposphere.

As we go up in the troposphere, the temperature drops. The reason for this is because the Earth's radiant energy gets less powerful as we go higher. The Earth's radiant energy is how the surface of the Earth releases energy that it absorbs from the sun. It mostly releases energy in the form of infrared rays.

The second layer of the atmosphere is the stratosphere. The stratosphere extends from 11 ~ 50 kilometers. In the stratosphere, convection does not occur because it is already hotter than the troposphere below it. This makes it the most stable layer, and airplanes travel within it because of it.

The stratosphere also contains the ozone layer, which absorbs UV rays and protects Earth by doing so. The ozone layer is located at 20 ~ 30 kilometers in the sky.

In the stratosphere, the temperature increases along with the altitude. This is due to the heat coming from the ozone layer as it absorbs UV rays.

The third layer is called the mesosphere. The mesosphere is located from 50 kilometers in the sky to 80 kilometers.

In the mesosphere, convection happens again because while the hot stratosphere is located underneath it, the mesosphere is comparatively colder because of its distance from both Earth's radiant energy and the ozone layer.

But, while convection happens, because there is very little water vapor in the mesosphere, meteorological phenomena do not happen. Also, meteors can also be seen in the mesosphere.

The last layer of the atmosphere is the thermosphere. The thermosphere stretches from 80 kilometers to about 1,000 kilometers, which is approximately where the atmosphere ends and space begins.

The air is very thin in the mesosphere and because of the less air, the differences in the temperatures of day and night are widely different. Auroras are also observed. The mesosphere is used as the route for satellites.

In the mesosphere, the temperature rises with

altitude because, while it is too far from both earth's radiant energy and the ozone layer, the proximity to the sun increases.

So, we can conclude that Hermes would only be able to go up to the troposphere, due to the other layers of the atmosphere being too thin to breathe in.

Chapter Seventeen :

Dionysus

The wine of Dionysus,

When the weary cares of men

Leave every heart.

We travel to a land that never was.

The poor grow rich, the rich grow great of heart.

All-conquering are the shafts made from the Vine.

02

Dionysus was the son of Zeus and the mortal Semele. Semele was the daughter of Cadmus and Harmonia of Thebes. Semele's relationship with Zeus ignited the fury of Hera and she sought revenge against Semele.

Hera knew she could not be directly involved in Semele's demise, for fear of being the target of Zeus's wrath. So, when searching for a way for revenge, Hera decided to use trickery against Semele instead.

The queen of Olympus appeared to Semele in the form of an old crone. Hera, in her disguise, pretended to befriend Semele. When Semele finally confided to Hera that the father of her child was Zeus, Hera was enraged.

However, she pushed her anger away and pretended

to be shocked and concerned for Semele. Hera planted the seeds of doubt against Zeus and Semele's belief in his true identity.

As a way to find out if Semele's lover was indeed Zeus and not a rich trickster, Hera suggested swearing him to an oath on the river Styx and asking him to appear to her in the form he appeared to his wife, Hera. Hera fully knew that a mortal could not gaze upon the true form of a god and used this fact to her advantage.

When Zeus came to visit Semele, she took Hera's ill meant advice and first swore Zeus on the river Styx to give her whatever she wanted. Zeus, thinking nothing of it, complied with her request easily.

When Semele asked him to appear to her in the form he appeared to Hera, however, Zeus, realizing the folly of his quick agreement to his oath, begged Semele to change her mind and to ask for something else, for laying eyes upon a god's form as they were in Olympus would kill the strongest of mortals.

However, Semele would not budge in her request, thinking that if the person she thought was Zeus could or would not comply, he was not the god he had meant her to believe he was.

Because Zeus was bound to complete the oath which he had sworn so foolishly on the river Styx, he had no other choice but to fulfill Semele's wish of him. Summoning only the lightest of thunderclouds and the weakest of his lightning bolts, Zeus appeared in front of Semele in his true godly form.

However weak Zeus's godly form was, it was still too much for Semele's eyes. Her soul fell to Hades in an instant, leaving only her body behind. Zeus, seeing that the baby inside of Semele was still miraculously alive, took it out of its mother and cutting open the flesh of his thigh, sowed the baby inside his body.

The god's thigh acted as an incubator for the child, giving him nourishment, along with immortality that came from subsisting off of the god's ambrosia and nectar. When the child was finally born, Zeus named it Dionysus, Dios meaning 'of Zeus' combined with Nysa, where Dionysus was to be raised. In all, Dionysus meant the son of Zeus in Nysa.

Before he was named for his staying at Nysa, Dionysus was first given to his mother's sister Ino to be raised. Ino and her husband gave Dionysus a false identity, raising him as a girl in order to throw Hera off his scent.

However, no god can be fooled forever and when Hera found Dionysus's foster home, her rage was terrible to behold. She inflicted Ino and her husband with madness and forced them to murder their own children, making them see them as wild animals and cooking ingredients for a stew.

Hera would later pull this stunt once more with another one of Zeus's illegitimate children, Hercules, except that in the latter case, she would inflict Hercules himself with madness, dirtying his hands with the blood of his own family.

Dionysus only escaped the brunt of her wrath because he was sent out to do some shopping for Ino when Hera came. Hermes was summoned by Zeus and took Dionysus to safety at Mount Nysa, where Hera could not find him.

The young god spent the rest of his childhood at Nysa, being raised by nature spirits, nymphs, and satyrs. At

Nysa, Dionysus discovered a vine that none of the spirits could identify.

After experimenting with the vine, he found that it bore fruit, fruit that was incomparable in taste with any other. Dionysus discovered that when he stored the juice of the fruit, it became sour and drove the nymphs and satyrs temporarily mad with its incredible taste. Dionysus had invented wine.

However, his childhood did not last long because once again, Hera had found out where Dionysus was staying and he was forced to be on the move again. His travels guided him around the world.

When he traveled to the city of Thebes, Dionysus was not recognized and dismissed by the king, Pentheus. Pentheus, also thinking that Dionysus was just another mad mortal, forbade the women of his city from joining his revelries.

In revenge for his actions against him, Dionysus caused all of the women to have an urge to sneak out of the city and join him in his revelry. After this had happened for a few days, Pentheus, hearing that his mother was also going out to join the revelry in secret, followed her to see what the party was like for himself.

When he arrived at the revelry, Dionysus made the tree, from which Pentheus was spying on the party, crack and fall to the ground. Pentheus was discovered by all of the drunk partygoers.

In their drunkenness, the women of Thebes did not recognize Pentheus as their king and instead mistook him for a wild animal. Converging on him like a horde of bees attracted to honey, they tore him apart.

It was only when the deed was done and the women were heading back to their homes with pieces of the wild animal to make breakfast that Dionysus released them from their drunken state of mind.

In another myth, Dionysus traveled to Attica to celebrate the new king of Athens. A man named Icarius warmly welcomed Dionysus who in gratitude, taught him the art of making wine.

Icarius was eager to share the god's kindness with his townspeople and went to the locals in order to share it with them. The locals, in their ignorance of what effect wine had on the state of mind, thought they had been poisoned and ended up murdering Icarius as a result.

When the sun rose on the next day and the locals woke up and realized what they had done, they buried Icarius. But Dionysus was not to be fooled. Taking pity on Icarius's daughter not knowing her father's demise, he led her dog to unearth her father's body so she could find it.

Upon seeing her father's corpse, Icarius's daughter despaired so much that she hung herself from a tree.

In his anger in place for Icarius and his daughter, Dionysus cursed the land of Athens to be barren and devoid of any water. He also afflicted the Athenian women with a madness which caused them to hang themselves, just like Icarus's daughter had.

It was only when the Athenians had prayed to the sun god, Apollo, for his advice when Dionysus accepted holding a festival in honor of Icarus and his daughter every year. The Athenians, not seeing another way out of their mess, agreed.

In another myth, Dionysus was captured by some Tyrrhenian pirates who promised him to give him a ride to Naxos but instead captured him as a hostage and they planned to sell him for a ransom.

Although Dionysus had told the truth about his father being the god Zeus when the pirates had asked, only the navigator of the ship had seemed uneasy and voted to let him go free, while the other pirates just thought Dionysus was lying to them.

Dionysus inflicted them with madness, making them see lions and panthers on the deck of the ship and jumping overboard to escape the phantom terrors. As they jumped into the sea, Dionysus turned them into dolphins.

One last final myth about Dionysus is the myth of the Golden Touch. When one of Dionysus's satyr friends, a satyr named Silenus, had gone missing, Dionysus looked everywhere to find him.

It turned out that Silenus had been living in the lap of luxury when king Midas had found the satyr and given him refuge from the streets in his own palace. Dionysus, pleased with the king's actions, granted the king a wish.

Midas, showcasing the greed that we humans are known for, decided on the Golden Touch, the power to turn everything he touched into pure gold. Dionysus granted his wish.

At first, for Midas, it was a dream come true. Every single thing that he touched, be it leaves or flowers or even whole trees turned into shiny pure gold.

Then, Midas started to suffer the consequences of his wish. If everything that he touched turned to gold, he could not eat food, because it turned into gold in his mouth. He could not even drink water or wine.

The final straw was when Midas, forgetting about his wish, gave his daughter a hug and she turned to gold under his very fingers.

Midas went back to Dionysus, begging him for a way to remove the Golden Touch. The god then told him that if he ran his hands over fresh, running water, he would lose the gift.

Midas easily made the decision and everything he transformed into gold turned back into their original forms. The leaves, flowers, trees, food, and most important of all, his daughter came back.

Amongst the many myths of Dionysus, the myth

that interested me most was the myth of King Midas and his golden touch. Could we, with our advanced future technology, make gold out of something that is originally not?

The short answer is that we can. But, like all the other topics we discussed, we need to observe the basic scientific facts at work. In this particular scenario, it is the structure of the atom.

The atom is made up of mainly two parts, which are the electron and the atomic nucleus. The electron has barely any mass at all, so little that even scientists exclude it from their calculations. The electron has a negative charge.

On the other hand, the atomic nucleus has most of the mass of the atom itself. It is made up of tiny particles called protons and neutrons, and they too are made up of even smaller particles, such as quarks.

The protons have a positive charge, while the neutrons don't have any charge at all, either positive or negative. The neutrons, while not having any electrical energy, have mass, just like their proton counterparts. This is the basic structure of the atom.

Now, if we want to look deeper, we can study the atomic models that were developed in the past, and how we, as our discoveries progressed, came to improve our previous models.

The people to truly first think about what the universe is composed of were the ancient Greeks. The scientists of Greece could be divided into believers of two widely different theories.

The first theory was one of the famous philosophers, Aristotle. Aristotle believed that the universe was made up of four core elements that he believed were central to reality. They were fire, air, water, and earth.

The second theory was the one of the Greek philosopher Democritus. Democritus believed that the universe was made up of small particles, called atoms, derived from the Greek word atomos, meaning uncuttable.

Later, Democritus's atomic model would be proven as truth while Aristotle's would be proven false, all due to the scientist Lavoisier. The French scientist, Lavoisier, conducted an experiment that could separate water into its basic elements, hydrogen, and oxygen.

Lavoisier would run water through a superheated steel pipe. Whilst flowing through the pipe, the water would dissolve, forming into hydrogen and oxygen. The highly reactive oxygen would react with the steel, forming iron oxide.

The hydrogen would travel through the pipe, be cooled down, and collected in a bottle of water. Lavoisier knew that the oxygen from the water had reacted with the steel pipe because, after his experiment, the weight of the steel increased.

By conducting this experiment, Lavoisier disproved Aristotle's theory of his four elements, fire, air, water, and earth by separating one of the aforementioned elements, water, into two different materials.

The first atomic model to be presented was John Dalton's atomic model. John Dalton, who was a scientist, improved on the old Greek scientist Democritus's theory. Dalton, like his predecessor, claimed that all matter is composed of very small things called atoms. In Dalton's theory, he claims four different truths.

- All matter is made of atoms and these atoms are

indivisible and indestructible

- All atoms of a choice element are identical in mass and properties
- Compounds are formed by a combination of two or more different types of elements
- A chemical reaction is a rearrangement of atoms

However, Dalton's theory is partially wrong. With the discovery of isotopes (atoms of the same element that have a difference in mass due to a difference in the number of neutrons), part of his theory was proven to be false.

What we can learn from this is that no one theory is absolute, and will never be. The best that we can do is to improve on the present model with new data given and discovered.

The second atomic model was Thompson's model. The scientist called Thompson discovered the electron and worked it into his model, which was given the name of the plum pudding model.

Thompson used a cathode ray tube with his

experiment for discovering the electron. Because cathode rays are negatively charged, and with several of his experiments, such as determining the electrical positivity with magnetic fields and testing its mass with a small turbine, Thompson concluded that all atoms are made up of very small negatively charged particles. Thus, the electron was born.

Thompson's plum pudding model stated that in positively charged particles, those particles with a negative charge were embedded into the positively charged one, like raisins in a pudding.

After Thompson's plum pudding model, the third model presented was Rutherford's model. The scientist conducted a revolutionary experiment, known as the gold foil experiment.

In the gold foil experiment, Rutherford beamed alpha particles (the nucleus of a helium atom) at a very thin gold foil. While most of the particles beamed at the foil would just go straight through it, a very small number of particles bounced off at odd angles or just curved to the side.

Because the alpha particle contained a positive

charge, Rutherford concluded that because of the curved particles, there must be some sort of positively charged particle making up the atom.

He also thought that because some of the alpha particles bounced off at odd angles, this meant that the positively charged atom had a very high mass. But, because most of the alpha particles missed it, it would also be very small.

We then move on to the fourth model of the atom: Bohr's model. Bohr's model was the basis for the model of the atom that we have today. Bohr's model describes the structure of the atom as the nucleus surrounded by electrons in circular orbits.

The electrons in these circular orbits could release energy by moving to an orbit closer to the nucleus, which would make the attraction between the two particles stronger so that it would not require much energy.

However, electrons could also move away from the nucleus. This would mean that the attraction becomes less powerful, making the electron unstable. In order to move to a further away orbit, electrons would need to absorb energy.

However, while Bohr's model would explain the emission spectrum of the atoms with only one electron, such as hydrogen, it did not work so well with the atoms containing more than one electron.

And so finally we come to the atom model of today: the electron cloud model. This model was calculated, and therefore contains some quantum mechanics information. The most important thing to focus on is Erwin Schrodinger's uncertainty principle.

Schrodinger states that an electron cannot be both observed in its location and its energy. With the very act of observation, one of the values would change every time. Therefore, we can only observe one at a time.

In the electron cloud model, it states that electrons are not moving around the nucleus on special, predetermined paths, but rather in a certain three-dimensional space called an orbital, which in itself is only the probability of an electron existing in that very space.

If we look at the quantum numbers that describe these orbitals, there are four of them. The first is the principal quantum number, represented with the letter n. The principal quantum number describes the energy level of the electron. The further away the electron is from the nucleus, the more unstable, and therefore more energy than the electron has.

The principal quantum number is only made up of positive number values. Also, it may be described with letters. For example, the first energy level (the level closest to the nucleus) is 1 and is marked with the letter K. With the second energy level it is L and so on.

Then we come to the angular momentum quantum number. This number describes the basic shape of the whole orbital and is represented with the letter l. However, the angular momentum quantum number is limited by the value of the principal quantum number.

The value of the angular momentum quantum number can only exist between the values of 0 and (n-1). Like with the principal quantum numbers, the angular momentum quantum numbers can also be written with letters. If the value is 0, the corresponding letter is 's', if the value is 1 it is 'p', and so on.

So, for example, if the principal quantum number is 2, then the angular momentum quantum number could be 0 or 1. Therefore, with the L, there is both a 's' and a 'p' in existence.

The third quantum number is the magnetic quantum number, or sometimes known as ml. The magnetic quantum number is fairly simple compared to the former two. It describes how different shapes of orbitals are shaped in space.

It is limited by the angular momentum quantum number and its value can only exist between the positive and negative values of the angular momentum quantum number.

For example, if l=1, m= -1, 0, 1 and therefore would exist three different ways in space.

The last and final quantum number is the spin quantum number. This value describes if the electron is spinning in a clockwise pattern or a counter-clockwise pattern. The value of the clockwise pattern is +1/2 while the value of the counterclockwise pattern is -1/2.

Electrons are also placed in configuration due to

three different rules. The first is the Aufbau principle. It states that electrons must always occupy the lowest energy orbital available and work their way up.

The second is Hund's rule. This states that electrons with the same spin quantum number as each other must occupy each orbital in a different sublevel before they pair up with an electron with a spin value different than theirs.

Lastly, we get the Pauli exclusion principle, which states that two electrons with the same spin value cannot occupy the same orbital and that a maximum of two electrons can occupy a single orbital.

Now if we go back to making gold out of mundane atoms, I must warn you that while this is a philosopher's stone come to life, the process is very expensive and will yield no profit.

We know that an atom has protons and neutrons in its nucleus and electrons. We also know that atoms change their properties with the difference between these three values.

What we can do to make gold is to match the number of protons, neutrons, and electrons in a single gold atom with the atom trying to turn into gold. With all of the requirements met, the atom would no doubt act with the properties of real gold.

/ Conclusion /

Greek mythology has a lot of meanings. It was the ancient Greeks' own way of explaining the world and its natural occurrences in their own meaningful and logical way. But the term "Greek mythology" all means different things to us. For some it may be a sanctuary to escape to when times are hard, for others, it may be a world to stretch their imagination to the limits, or even a way to socialize and connect with other people.

By observing the multitude of gods, goddesses, monsters, and mortals, we can therefore discover our true selves in the act of beholding the myths. Because of this, while the science of the ancient world is disproven by the science of today, it is immortal nonetheless. It continues and lives on in all of us.

Praise for <The Science of Olympus>

It is such a pleasure to recommend Joonyoung's book, 'The Science of Olympus.' The first time I had the pleasure of meeting with Joonyoung goes back to my days in Washington, DC as Korea's Ambassador there. At a dinner with Joonyoung's family, I raised a question that a grown-up often raises with a 10 year old boy; 'How do you like your school?' It was a casual question, but Joonyoung's answer was nothing but. He began to talk about his favorite class, the history of the Commonwealth of Virginia, which was so thorough and inspired that I was quickly convinced that Joonyoung was a boy of exceptional intellectual curiosity and capacity.

As for 'The Science of Olympus,' I was at first intrigued by the title of the book and wondered what Greek mythology has to do with science. Reading through the book, I was truly captivated by the fresh perspective Joonyoung brings to his reading and interpretation of Greek mythology. In the Introducti ontothebook,Joonyoungwrote;'Iwrote this book with the intention of telling the science of a world long past, lost to the sands of time, along with the science of today." Joonyoung brilliantly succeeded to fulfill the intention, and I full-heartedly recommend his book.

- Ahn Ho-young, former South Korean Ambassador to the United States

Myths are structured around real events and often delve into the matters of heroes or salvation from a godly figure. In this way, myths help us humans lead out a meaningful life. Because the Greek myths are legends based on local beliefs and historical events, they have led to inspire the artistic works of today, along with influencing the thoughts of philosophers and historians.

Also, the symbolism around the variety of gods, heroes, and monsters and their actions featured in the Greek myths have continually been revived and reinterpreted with different meanings. However, they all have a special meaning to each of us based on our interests and philosophical beliefs.

Joonyoung has held a special interest in the Greek myths and what ancient science they hold within them, and he has connected this interest with modern science. Myths and science are both hard and difficult areas of knowledge, but by looping these two subjects together, Joonyoung has shown his prowess and wit. I firmly believe that Joonyoung's passion for his interests will go on to achieve many great dreams, and I applaud him.

- Kim Mi-ok, principal of Mogun Middle School

The Principal Gods

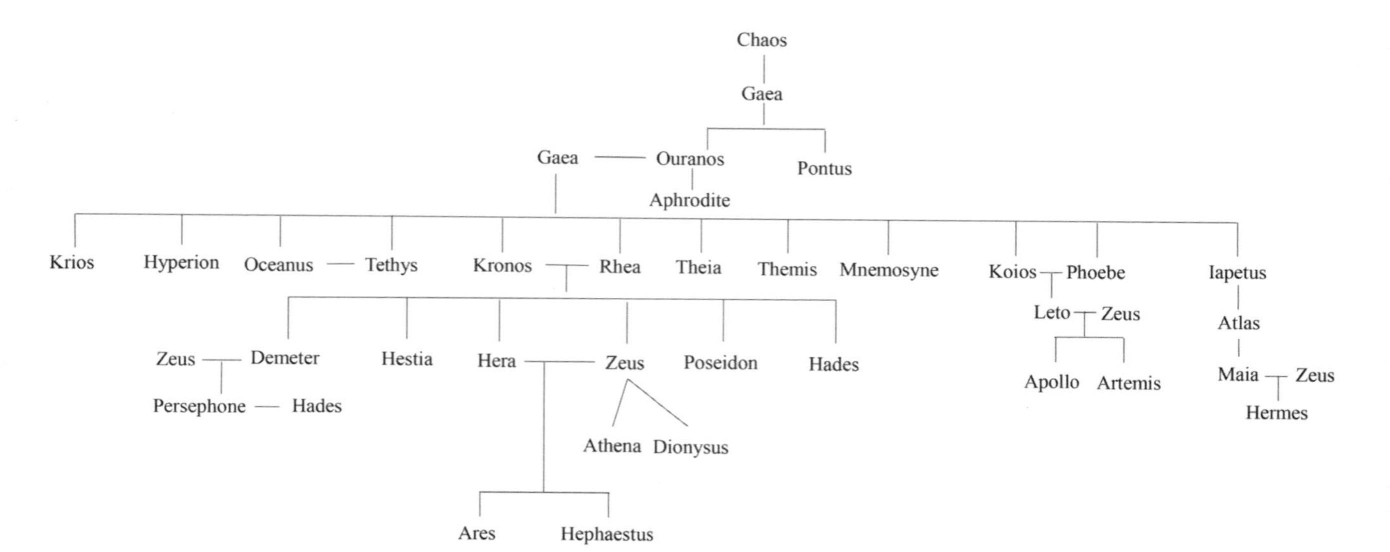

THE
SCIENCE
OF
OLYMPUS

초판 1쇄 발행 2021. 11. 10.

지은이 장준영 (Jay Joonyoung Chang)
펴낸이 김병호
편집진행 김수현 | **디자인** 최유리

펴낸곳 주식회사 바른북스
등록 2019년 4월 3일 제2019-000040호
주소 서울시 성동구 연무장5길 9-16, 301호 (성수동2가, 블루스톤타워)
대표전화 070-7857-9719 **경영지원** 02-3409-9719 **팩스** 070-7610-9820
이메일 barunbooks21@naver.com **원고투고** barunbooks21@naver.com
홈페이지 www.barunbooks.com **공식 블로그** blog.naver.com/barunbooks7
공식 포스트 post.naver.com/barunbooks7 **페이스북** facebook.com/barunbooks7

· 책값은 뒤표지에 있습니다. **ISBN** 979-11-6545-535-4 03400